Green Energy and Technology

Climate change, environmental impact and the limited natural resources urge scientific research and novel technical solutions. The monograph series Green Energy and Technology serves as a publishing platform for scientific and technological approaches to "green"—i.e. environmentally friendly and sustainable—technologies. While a focus lies on energy and power supply, it also covers "green" solutions in industrial engineering and engineering design. Green Energy and Technology addresses researchers, advanced students, technical consultants as well as decision makers in industries and politics. Hence, the level of presentation spans from instructional to highly technical.

Indexed in Scopus.

Indexed in Ei Compendex.

N. Domingo · S. Senarathne · Suzanne Wilkinson · Srinath Perera
Editors

Global Net Zero Carbon Practices in Construction

Springer

Editors
N. Domingo
School of Built Environment
Massey University
Auckland, New Zealand

S. Senarathne
School of Engineering, Design and Built Environment
Western Sydney University
Sydney, WA, Australia

Suzanne Wilkinson
School of Built Environment
Massey University
Auckland, New Zealand

Srinath Perera
School of Engineering, Design and Built Environment
Western Sydney University
Sydney, NSW, Australia

ISSN 1865-3529 ISSN 1865-3537 (electronic)
Green Energy and Technology
ISBN 978-981-96-3855-0 ISBN 978-981-96-3853-6 (eBook)
https://doi.org/10.1007/978-981-96-3853-6

This Springer imprint is published by the registered company Springer Nature Singapore Pte Ltd.
The registered company address is: 152 Beach Road, #21-01/04 Gateway East, Singapore 189721, Singapore

Contents

Introduction

N. Domingo and S. Senarathne

1 Introduction

As global concerns over climate change intensify, achieving net zero carbon emissions has become a critical target across industries, with construction positioned as a key player in this transformative journey. The construction industry is one of the major contributors to global greenhouse gas emissions [1]. According to a 2019 report by the World Green Building Council, buildings account for 39% of global energy-related carbon emissions, with 28% stemming from operational emissions and the remaining 11% from embodied emissions, those associated with materials and construction processes [2]. In 2021, global carbon emissions from building operations alone saw a 5% increase over 2020, underscoring the urgency of reducing emissions in this sector [3].

Global temperatures correlate directly with the cumulative carbon emissions released into the atmosphere, making emission reductions vital to preventing severe climate impacts [4]. The 2015 United Nations Paris Agreement set forth an international commitment to limit global temperature increases to below 2 °C, ideally 1.5 °C above pre-industrial levels, a target agreed upon by world leaders [5]. In light of this mandate, designing and constructing net zero carbon buildings and structures that generate as much renewable energy as they consume, with minimal or no carbon emissions over their entire life cycle has become a crucial priority.

The path to net zero carbon in construction is complex, involving the integration of sustainable materials, renewable energy sources, and energy-efficient designs, while also addressing demands for durability, functionality, and economic viability.

N. Domingo (✉)
School of Built Environment, Massey University, Auckland, New Zealand
e-mail: n.d.domingo@massey.ac.nz

S. Senarathne
School of Engineering, Design and Built Environment, Western Sydney University, Sydney, Australia

N. Domingo et al. (eds.), *Global Net Zero Carbon Practices in Construction*, Green Energy and Technology, https://doi.org/10.1007/978-981-96-3853-6_1

Achieving net zero requires a shift from traditional practices to an interdisciplinary approach, bringing together architects, engineers, contractors, and policymakers to develop innovative, practical, and economically feasible solutions. Adopting net zero practices not only addresses the urgent need for climate action but also contributes to healthier, more resilient built environments that can withstand climate extremes while reducing long-term operating costs.

This book, Global Net Zero Carbon Practices in Construction, provides a comprehensive overview of the essential strategies and practices emerging globally. It brings together leading-edge research, case studies, and industry insights, exploring diverse approaches and methodologies for achieving net zero carbon goals. This volume is a valuable resource for practitioners, academics, policymakers, and students, guiding them through the complexities of sustainable building design and construction. Each chapter highlights successful initiatives, technical innovations, and the socioeconomic implications of shifting to a net zero carbon framework, offering a global perspective on both challenges and opportunities across different regional contexts.

As the world collectively pursues the Paris Agreement targets and broader climate goals, this book underscores the importance of net zero carbon practices as not only a means of environmental stewardship but as an imperative for future-proofing the built environment. This introduction chapter sets the tone for the entire volume by emphasising the importance of reducing carbon emissions within the construction sector. Through a thought-provoking overview, readers are prepared to engage with subsequent chapters, which will unfold innovative strategies, best practices, and case studies aimed at creating a more sustainable and environmentally responsible construction industry worldwide. The following sections briefly describe the contents of the subsequent chapters.

Chapter 2 establishes a clear understanding of the concept of net zero carbon within the construction domain, offering a nuanced definition that sets the stage for subsequent discussions. The chapter provides a valuable historical context, through a systematic literature review. First, through a bibliometric analysis, it identifies research publication trends, keyword occurrences, and global spread country-wise, over the last two decades. Next, after shortlisting seminal papers in the field, a content analysis was undertaken to trace the evolution of carbon reduction efforts in construction considering three eras: Kyoto Protocol Era, Pre-Paris Agreement Era and, Post-Paris Agreement Era. Chapter highlighted key global initiatives during these eras that have shaped the industry's approach towards a more environmentally conscious future.

Chapter 3 addresses the role of the buildings and construction industry in achieving net zero carbon (NZC) goals. Focusing on NZC regulations and policies in nine countries USA, Japan, Germany, South Korea, Canada, China, India, Indonesia, and Saudi Arabia, the chapter examines each country's regulatory landscape, compliance challenges, and opportunities for advancing carbon neutrality. It presents action plans for these nations and highlights differences in approach between developed and developing countries in the pursuit of carbon neutrality. The findings underscore the importance of NZC regulations in the top carbon-emitting countries, offering insights for effective policy development and compliance. The chapter also discusses the need for collaboration among governments, policymakers, industry professionals, and researchers to navigate challenges and foster progress on the path to carbon neutrality.

Chapter 4 aims to provide insights into stakeholder perspectives and presents success stories that showcase effective collaboration in achieving sustainable construction goals. Additionally, the chapter addresses the imperative of preparing the workforce through education and training programmes, recognising the significance of skilled professionals in implementing and maintaining environmentally conscious construction practices on a global scale.

Chapter 5 explores innovative solutions available for accurately tracking carbon emissions throughout the construction lifecycle using a systematic literature review. It highlights the transformative role of technology in promoting low-carbon practices within the industry and showcases various tools that contribute to enhanced efficiency and sustainability. It outlines the historical progression and provides a clear direction for future research, guiding scholars in exploring the practical applications of these technologies for reducing carbon emissions within the industry.

Chapter 6 delves into the pressing issue of carbon emissions in the construction industry, which stem from both operational carbon and embodied carbon. Recognising the urgency of reducing these emissions to meet global climate goals, the chapter provides an in-depth examination of strategies to minimise operational carbon through energy optimisation and embodied carbon through sustainable materials and methods. The content is grounded in a critical review of contemporary research, offering insights into passive and active design strategies, material innovations, and construction methods that contribute to carbon reduction. It highlights the role of tools like life cycle assessment (LCA) and environmental product declarations (EPDs) in assessing carbon impacts, along with the application of digital tools such as building information modelling (BIM) for enhancing energy efficiency. Through this synthesis, the chapter equips industry stakeholders including designers, engineers, and policymakers with evidence-based approaches to lower carbon emissions and informs sustainability standards that support the industry's transition towards carbon neutrality.

Chapter 7 focuses on supply chain and material procurement for low-carbon construction. Through a systematic literature review, the chapter identifies highest impactful construction materials as steel, concrete, and aluminium. Then, it explores strategies to minimise carbon impact during cradle-to-gate and gate-to-site by sourcing eco-friendly materials and optimising procurement processes. Simultaneously, the chapter addresses on-site low-carbon practices, providing insights into construction methodologies that reduce emissions during the building process. By addressing both the upstream and on-site components of construction, this chapter equips readers with practical knowledge to foster sustainable practices towards net zero carbon in the construction industry.

Chapter 8 explores the role of maintenance and refurbishment in reducing greenhouse gas emissions from existing buildings. It underscores the importance of proactive maintenance strategies: preventive, corrective, and predictive, that can extend building lifespans, enhance energy efficiency, and decrease the need for resource-intensive renovations. Focusing on building refurbishment, the chapter discusses energy-efficient retrofits, such as improved insulation, windows, lighting, and HVAC systems, as well as the integration of renewable energy and smart building technologies to lower operational emissions. It also covers greenhouse gas quantification, referencing the greenhouse gas protocol, and addresses performance benchmarking for emissions comparison. Additionally, the chapter presents methods for evaluating the environmental and economic impacts of retrofits, featuring an empirical case study that demonstrates these evaluations in practice. By examining these strategies, this chapter provides readers with insights into making the built environment more sustainable through effective maintenance and refurbishment.

Chapter 9 explores waste management strategies and circular economy principles within the construction industry, focusing on responsible material disposal and recycling. It introduces deconstruction techniques aimed at material recovery, highlighting innovative approaches to dismantling structures for salvaging and reuse. To assess the alignment between research efforts and policy objectives on waste management, the chapter presents a systematic literature review of 126 studies from 2019 to 2023, covering themes like management practices, recycling and upcycling, circular economy, stakeholder behaviour, and technological advances. The chapter underscores the need for targeted research, providing a roadmap for researchers and industry professionals to advance CDW management and support policy goals.

Chapter 10 addresses the pressing challenges of urban sustainability, focusing on how urban regeneration can mitigate climate impacts and enhance community resilience. With rapid urbanisation contributing to pollution, congestion, and loss of green spaces, the chapter explores the role of blue-green infrastructure integrated open and green spaces offering ecological, economic, social, and aesthetic benefits for urban areas. Using Istanbul as a case study, the chapter examines innovative approaches to urban regeneration, specifically through repurposing underutilised sites like gas stations in line with acupuncture urban planning principles. It introduces a methodology employing key performance indicators (KPIs), SNTool, and GIS-based impact assessment to guide the transformation of these sites into green spaces that improve environmental quality and urban livability. The case study demonstrates how converting gas stations into green infrastructure can enhance Istanbul's urban landscape, addressing urbanisation and climate challenges effectively.

Chapter 11 synthesises the key findings and insights from the book, highlighting the advancements, challenges, and practical approaches in implementing net zero carbon (NZC) practices within the construction industry. It discusses the main barriers to NZC adoption, such as regulatory hurdles, financial constraints, and gaps in stakeholder engagement, while also identifying opportunities for innovation through technology, policy alignment, and collaborative efforts across industry and government. The chapter also outlines a forward-looking perspective on the future of NZC practices, emphasising the need for continued research, stronger policies, and investment in low-carbon technologies.

References

1. Akan MÖA, Dhavale DG, Sarkis J (2017) Greenhouse gas emissions in the construction industry: an analysis and evaluation of a concrete supply chain. J Clean Prod 167:1195–1207
2. World Green Building Council (2019) Global status report for buildings and construction
3. UN Environment Programme (2022) Annual report 2022
4. Rogelj J et al (2015) Zero emission targets as long-term global goals for climate protection. Environ Res Lett 10(10):105007
5. United Nations (2015) Paris Agreement. United Nations, Paris, France

Introduction to Net Zero Carbon in Construction

S. Senaratne and M. Nadeeshani

Abstract Society is facing significant problems due to climate change. The construction industry is a key contributor to greenhouse gas emissions. Buildings alone account for 39% of global energy-related carbon emissions. Net zero carbon emerged as an essential concept in reducing greenhouse gas emissions in the construction sector. This chapter aims to establish a clear understanding of the concept of net zero carbon within the construction domain, underscore the vital role of low-carbon practices, and offer valuable historical context of net zero carbon buildings. A systematic literature review was conducted, and the study used a mixed-method research approach. First, a bibliometric analysis was conducted on 413 publications retrieved from the Scopus database. The analysis revealed that research on net zero carbon in the construction sector began in 2004. A gradual increase in publications was evident after 2019, and the UK, China, and the USA were the dominant contributors to the current literature. Second, a qualitative content analysis was conducted on carefully selected 38 articles. The results highlighted two key milestones which shaped the industry's approach towards a net zero future: the Kyoto Protocol and the Paris Agreement. During the Kyoto Protocol era, the main research themes included clean development mechanisms, net zero energy buildings, and carbon neutrality goals by countries. This was shifted towards net zero carbon, nationally determined contributions, net zero carbon targets, carbon reduction technologies/ policies, and roadmaps towards carbon neutrality, by the Paris Agreement era.

Keywords Net zero carbon · Net zero carbon buildings · Kyoto protocol · Paris agreement

---

S. Senaratne (✉) · M. Nadeeshani
Research Centre for Smart Modern Construction, School of Engineering, Design and Built Environment, Western Sydney University, Sydney, Australia
e-mail: s.senaratne@westernsydney.edu.au

M. Nadeeshani
e-mail: M.SammuHandi@westernsydney.edu.au

N. Domingo et al. (eds.), *Global Net Zero Carbon Practices in Construction*, Green Energy and Technology, https://doi.org/10.1007/978-981-96-3853-6_2

1 Introduction

Society is facing significant problems due to climate change. Greenhouse gas emissions (GHG) from human activities are causing climate change and global warming. Carbon dioxide (CO_2) is the main GHG accountable for these adverse effects, primarily generated by deforestation, industrial processes, and burning fossil fuels. These emissions disrupt the natural balance of the global ecosystem, contributing to rising temperatures, frequent and extreme weather events, and other environmental issues.

The construction industry significantly contributes to global GHG emissions. Buildings account for 39% of global energy-related carbon emissions. This includes 28% of operational and 11% of embodied carbon emissions [1]. The building sector is also responsible for 33% of electricity consumption in Europe in 2020 and the figure is expected to rise to 72% by 2050 [2]. As the latest global initiative to these pressing issues, the Paris Climate Agreement of 2015 was signed establishing targets to limit temperature increases below 2 °C or ideally 1.5 °C [3]. As a result, many countries are committing to achieve net zero emissions targets set by the Paris Agreement by mid-century.

The main sources of carbon emissions in the construction sector are the energy utilisation required for material production and transportation, construction activities, and building operations. Reducing these carbon emissions within the building life cycle can contribute significantly to mitigating climate change and global warming. Low-carbon practices have become the topics of current interest in reducing carbon emissions in the built environment. Low-carbon practices within the construction sector are mainly attributed to energy as buildings require a substantial amount of non-renewable energy to function [4]. Therefore, energy efficiency and renewable energy are considered key concerns in reducing operational carbon emissions [5]. Embodied carbon which refers to emissions generated throughout the lifecycle of materials and products, from extraction to disposal, is considered the other type of emission that needs to be reduced through low-carbon practices.

Carbon reduction efforts in the construction sector have evolved over time. Starting from low-carbon buildings which only attempt to reduce carbon emissions to the environment, the building sector is gradually transforming towards no carbon by strengthening the carbon reduction efforts, preventing possible carbon emissions, and balancing them with removals. Different concepts like low carbon and zero carbon have emerged over time to streamline carbon reduction efforts and direct the industry towards an emission-free future. However, the latest trend towards carbon neutrality in the built environment sector is net zero carbon buildings (NZCB).

The concept of 'net zero' in the construction sector has rapidly and recently become the topic of current interest. There are different terminologies are being used by various countries when referring to net zero emission targets. Several terms such as net zero emissions, net zero carbon, and net zero GHGs are used collectively and widely [6]. However, they all share a common goal of reducing or neutralising the environmental impact of buildings by low-carbon practices in the sector. A clear definition and understanding are useful to maintain the consistency of a concept. Therefore, this chapter aims to establish a clear understanding of the concept of net zero carbon within the construction domain, offering a nuanced definition that sets the stage for subsequent discussions. This chapter then provides a valuable historical context, tracing the evolution of carbon reduction efforts in the construction sector worldwide. Additionally, it highlights key global initiatives that have shaped the industry's approach towards a more environmentally conscious future.

2 Net Zero Carbon in Construction

The term 'net zero' is a word of balance. When global GHG emissions are cancelled out or removed in equal to already emitted GHG emissions, it is considered a net zero GHG emission. In addition to the balancing metric, this balancing system generally requires a timeframe to be regarded as a net zero, within which this balance must be achieved, such as annually, monthly, or over the lifecycle of a product [7]. The first appearance of the term 'net zero' is related to the fields of physics and physical chemistry and appeared in terms like net zero field mobility and net zero phase change in 1957 [8]. The first climate-related use of the term net zero was evident in 1991, in a study that explored biomass fuels to approach zero net CO_2 emissions [8, 9]. Gradually, the term net zero was more frequently used in climate activities and construction sector from the early 2000s due to the global focus towards sustainability and GHG reduction.

The usage of the term 'net zero' in the construction sector varies by different metrics of balance. There are plenty of different metrics of balance that have been used with the term 'net zero' in the academic literature. Mostly, the term net zero is combined with 'carbon' or 'energy' as the balancing metric in the building sector. According to Loveday et al. [8], when 'carbon' is considered the balancing metric, terms like 'net zero carbon buildings', 'net zero carbon homes', and 'net zero carbon emissions' were widely used in the literature. When 'energy' is considered the metric of balance, 'net zero energy building' is the most commonly used term in the building sector [8]. However, it is worth noting the connectivity and collective use of 'energy' and 'carbon' along with the net zero concepts in the building industry. When the energy consumption of a building is reduced, the associated GHG emissions are also decreased. Hence, increasing energy efficiency and reducing energy consumption are considered similar to reducing carbon emissions in a building [10]. This interconnectivity has led the different terms of energy (e.g. net zero energy, zero energy, embodied energy, operations energy, etc.) to appear collectively and widely with NZCBs.

A specific focus on carbon in the built environment sector reveals several evolved concepts and terminologies. Literature offers various concepts and classifications for carbon targets in the built environment. Initially started with a low-carbon concept, various carbon targets of the building sector have been set and defined under different names and concepts such as carbon negative, carbon neutral, climate positive, net zero carbon, and absolute zero carbon. Each concept contains and focuses on specific values and has set specific carbon emission goals from a sustainability viewpoint [2]. For example, carbon-neutral buildings aim to achieve no net release of GHGs in their life cycle [2, 11], while climate-positive buildings mainly aim at achieving net zero emissions from the operational phase of the building [12]. The emissions in other building phases will be kept minimised and responsibly offset. Carbon-negative buildings, on the other hand, go beyond carbon neutrality by compensating more for their carbon emissions into the atmosphere [2]. In counteracting the carbon footprint created by the construction industry, the carbon-negative buildings aim to achieve a negative carbon footprint by removing more CO_2 than they omitted during the building's lifecycle. Absolute zero carbon attempts to avoid carbon emissions in the operational and construction phases of a building by using renewable energy and zero emission supply chains [13]. It aims to make carbon emissions absolutely zero without even using carbon offsets. However, this concept is still at the conceptual level. Contrary to absolute zero carbon emissions, NZCBs allow a balance between carbon production and removal from the atmosphere by a building. The net carbon emission of NZCBs over its life cycle could be a zero or negative value [14]. Net zero is often considered a synonym for carbon neutrality. But, according to the Green Building Council of Australia (2023), net zero refers to a 'state' or 'end point', while carbon neutrality is more often referred to as a 'journey'. Table 1 presents an overview of each concept.

Several terms are used by different countries when discussing net zero carbon in the built environment. Various terms like net zero emissions, net zero carbon, and net zero GHGs are used interchangeably [6]. For instance, countries like Australia, the USA, and Canada have discussed reducing GHG emissions to achieve net zero targets set by the Paris Agreement. The USA and Canada specifically considered actions to be taken to reduce carbon and non-carbon GHG emissions [6]. An overview of the carbon-related terminology in construction is presented in Fig. 1.

Several researchers have also highlighted the lack of common and shared definitions for NZCBs. Therefore, various definitions for local context are used by various countries. Mostly, these definitions are published by local green building councils (GBCs) and some such definitions are presented in Table 2. This lack of a shared definition for NZCBs also leads the countries to use different terms to refer to NZCBs. Specially in New Zealand and Norway are using the terms 'net zero buildings' and 'zero carbon buildings' respectively to refer to net zero carbon in the built environment. However, the UK Green Building Council's definition is the clearest among the other definitions presented in Table 2. It includes the building life cycle as the timeframe to achieve net zero emissions, further mentioning the net zero carbon—construction (embodied emissions) and net zero carbon—operations (operational carbon emissions). It was originally provided definitions for net zero embodied and operational carbon emissions separately to arrive at the net zero whole life carbon emissions definition presented in Table 2. It is evident that the other definitions are mostly focused on the operational phase (the USA, and South Africa) and some definitions are unable to provide a specific timeframe to achieve net zero carbon emissions. In conclusion, many researchers have referred to the NZCBs as a state

Table 1 Overview of different carbon concepts in construction sector

Concept	Definition	Characteristics	References
Carbon neutral	Buildings that result in no net release of GHG emissions into the atmosphere, often using offsets	• No net release of GHG • Often use offsets • Align with 1.5–2 °C of warming	Tirelli and Besana [2], Causone et al. [11]
Climate positive	Buildings that have no net emissions of GHG from operations due to renewable energy use, minimise all other emissions and responsibly offset the remaining	• No net emission of GHG from operational phase • Minimise other emissions • Offset the remaining emissions responsibly • Align with 1.5 °C or less warming	Green Building Council Australia [12]
Carbon negative	Buildings that achieve carbon neutrality and go beyond to reverse negative impacts of development and construction	• No net emission of GHG • Embodied carbon is the main concern • Align with 1.5 °C or less warming	Birkeland [15], Tirelli and Besana [2]
Net zero carbon	Buildings which produce zero or negative carbon emissions over the life cycle through the use of renewable sources and offsets	• Zero or negative release of carbon emissions over the life cycle • Use offsets and renewable sources • Align with 1.5–2 °C of warming	UKGBC [14],
Absolute zero carbon	Carbon emissions are avoided in operational and construction phases by using renewable energy and zero emission supply chains, without using offsets	• Avoid emissions through zero supply chain and renewable energy • Currently at conceptual level	WLCN [13]

where "the amount of carbon emissions associated with a building's embodied and operational impacts over the life of the building, including its disposal are zero or negative", inspired by the UK definition.

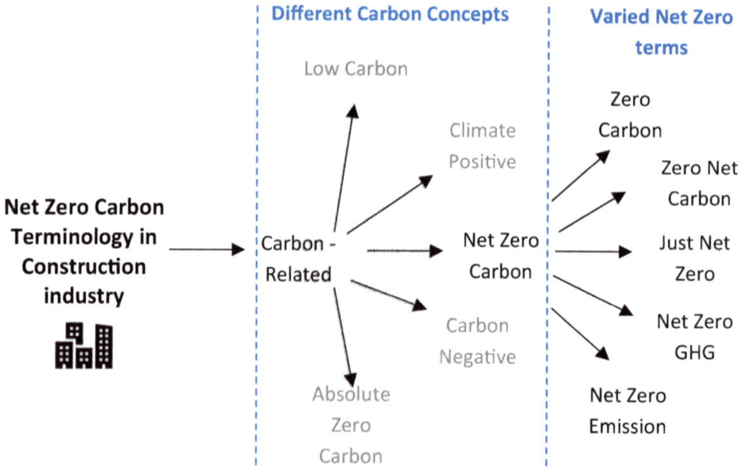

Fig. 1 Overview of carbon-related terminology in construction sector

Table 2 Different definitions for 'net zero carbon' in built environment by countries

Country	Definition	Reference
UK	"When the amount of carbon emissions associated with a building's embodied and operational impacts over the life of the building, including its disposal, are zero or negative"	UKGBC [14]
Australia	"Cutting greenhouse gas emissions to as close to zero as possible, with any remaining emissions re-absorbed from the atmosphere"	GBCA [16]
Ireland	"Total global warming potential for all cradle to grave life cycle stages are less than or equal to zero, where residual carbon is compensated for via renewable energy sources, or as a last resort carbon offsetting"	IGBC [17]
USA	"Highly efficient building that achieves a zero balance of carbon emissions emitted during operations"	USGBC [18]
South Africa	"A building that is highly energy-efficient, with the remaining energy requirement generated from renewable energy, preferably on-site or off-site where absolutely necessary"	GBCSA [19]

3 Research Methodology

The study conducted a systematic literature review to explore trends and key global initiatives towards NZCBs. The systematic literature review was carried out in two parts, a quantitative bibliometric analysis and a qualitative content analysis. The systematic review provides a valuable historical context, tracing the evolution of carbon reduction efforts in construction, and highlights key global initiatives that have shaped the industry's approach towards net zero built environment. Systematic identification and collation of literature available under specific research area is referred to as a systematic literature review [20]. The Preferred Reporting Items for

Systematic Reviews and Meta-Analyses (PRISMA) 2020 protocol, which aims to enhance the scientific quality and transparency of reviews, was used to conduct the systematic review [21, 22]. The clear methodological and analytical approach of the PRISMA protocol made it suitable for screening and evaluating information from large academic databases [23]. Figure 2 presents the developed PRISMA framework for the study.

The search string used to retrieve the publications on NZCBs in the construction industry was ('net zero carbon' OR 'net zero carbon') AND ('building' OR 'construction' OR 'built environment' OR 'hous*' OR 'home'). Amidst different types of carbon-related terms in the industry, the search was limited only to net zero carbon due to the wide use of this term and the specific focus towards net zero carbon in this chapter. The asterisk sign (*) in the search string was used to capture the variations in root words. For example, the term 'hous*' was used to capture variations

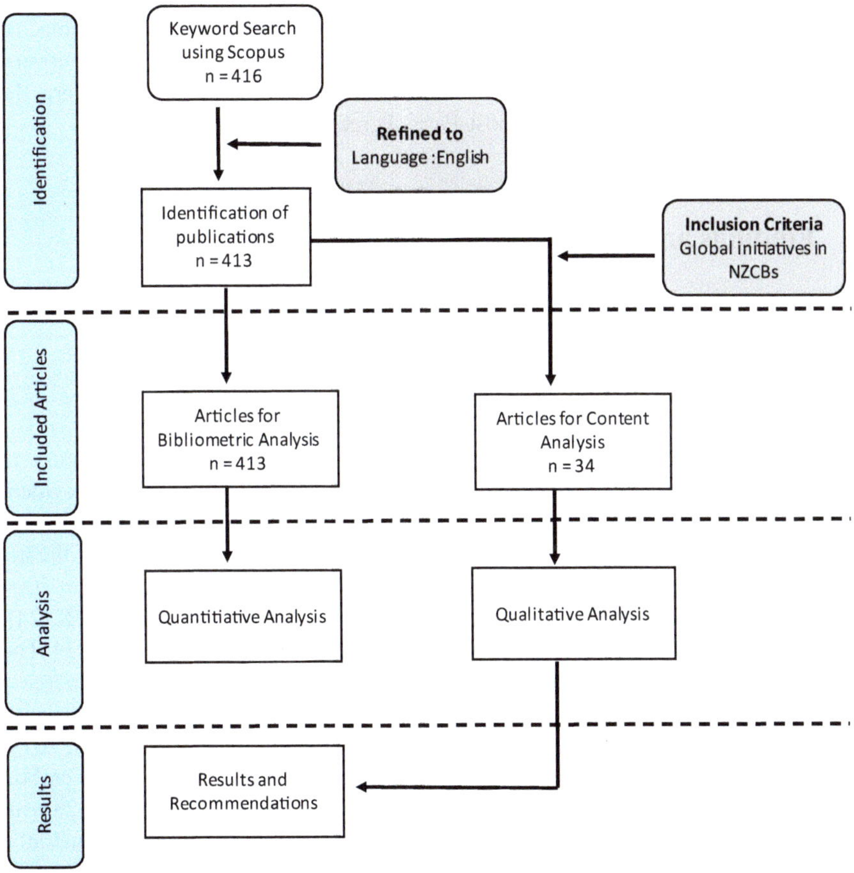

Fig. 2 Overview of research methodology

such as ‘house’, ‘housing’, ‘houses’, and alike. The literature retrieval was performed using the Scopus database as it covers an extensive range of scientific publications.

The initial search in the Scopus database resulted in retrieving 416 scientific publications, as of July 2024. When the published language was limited to English, the number of articles was reduced to 413. The bibliometric analysis was conducted on these 413 publications using software and manual analysis. Two main software were used in this study to conduct bibliometric analysis; VOS Viewer version 1.6.20, and Biblioshiny from R studio. The bibliometric analysis is presented using networks and maps such as keyword co-occurrence analysis and scientific production by countries network to enable visualisation and discovery of trends, patterns, and evolution of the NZCB research.

413 papers were further shortlisted using inclusion and exclusion criteria to yield the most relevant papers for qualitative analysis. The qualitative analysis was aimed at identifying the timeline and key global initiatives that shaped the industry towards a net zero carbon future. The papers that are irrelevant to this objective were excluded after carefully screening the content. This resulted in shortlisting 34 publications. At the final stage, these papers were allocated to sub-themes based on the timeframe and key global initiatives. The organised themes include the Kyoto Protocol era, the pre-Paris Agreement era, and the post-Paris Agreement implementations.

4 Findings and Discussion

4.1 Bibliometric Analysis

4.1.1 Trends of Research Towards NZCBs

Even though the ‘net zero’ term first appeared in the mid-twentieth century in academic literature, the first climate-related use of the term net zero was evident in 1991, in a study that explored biomass fuels to approach zero net CO_2 emissions [8, 9]. However, according to the current study, the first appearance of the net zero carbon concept in the construction sector appears to be Jones’s study in 2004 which examined the first commercially built carbon-neutral building in the UK [24]. Figure 3 presents the annual trend of publications on NZCBs from 2004 to 2024. The findings show a gradual increase in publications from 2004 onward. A slight increase in the research publications on NZCBs can be observed from 2012 compared to early 2000s levels. However, a significant rise in NZCB research can be seen from 2019, possibly due to the focused attention towards carbon neutrality and net zero carbon targets set by the Paris Agreement in 2015. The number reached its peak in 2023 with 122 publications. Since the current study is conducted in mid-2024, the illustrated number of publications in 2024 is only partial and does not present the final count for the year.

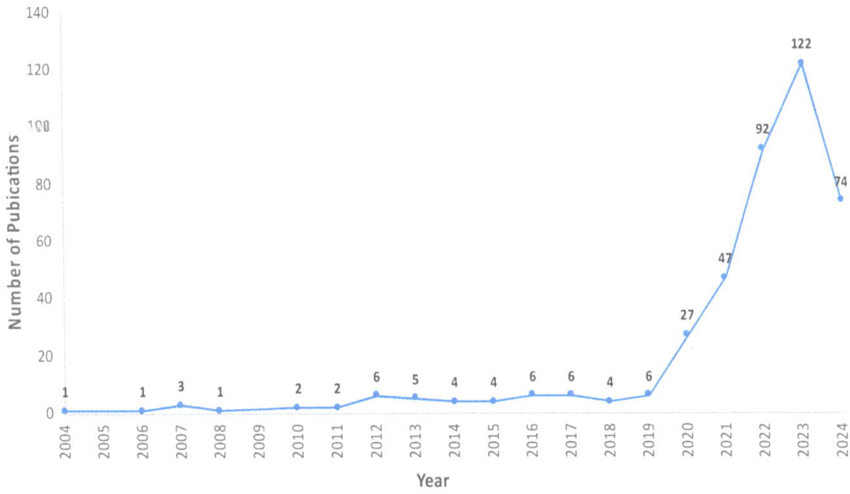

Fig. 3 Trend in research publication on NZCBs

4.1.2 Co-occurrence Network of Keywords

The theme of a study can be reflected by the keywords of a research study. When the keywords of a particular field of study are mapped together, it offers a clear presentation of the existing knowledge [25]. Also, the keyword's co-occurrence network illustrates the relationships between keywords. The co-occurrence network of keywords was created for the current study, using the VOS Viewer software. The network that was created using the author keywords is presented in Fig. 4.

The minimum number of keyword occurrences was set to 3 in VOS Viewer to generate a legible and optimum network. Out of 1337 keywords, 68 keywords met the threshold with 8 clusters, 297 links, and a total link strength of 409. Keywords with similar meanings, such as singular and plural forms, abbreviations, and variations in language use, were merged into single keywords using a thesaurus file. For example, 'building' and 'buildings', 'GHG emission' and 'greenhouse gas emission', and 'decarbonisation' and 'decarbonization' were merged. The size of each node in Fig. 4 represents the number of occurrences of the keyword, whereas the link strength of two connecting keywords indicates the number of articles they appeared together. The top five keywords identified are the life cycle assessment, net zero carbon, embodied carbon, climate change, and carbon emissions. It was also noted that the keyword 'net zero carbon' is highly appeared (strongly linked) with life cycle assessment, embodied carbon, climate change, buildings, and sustainability, respectively.

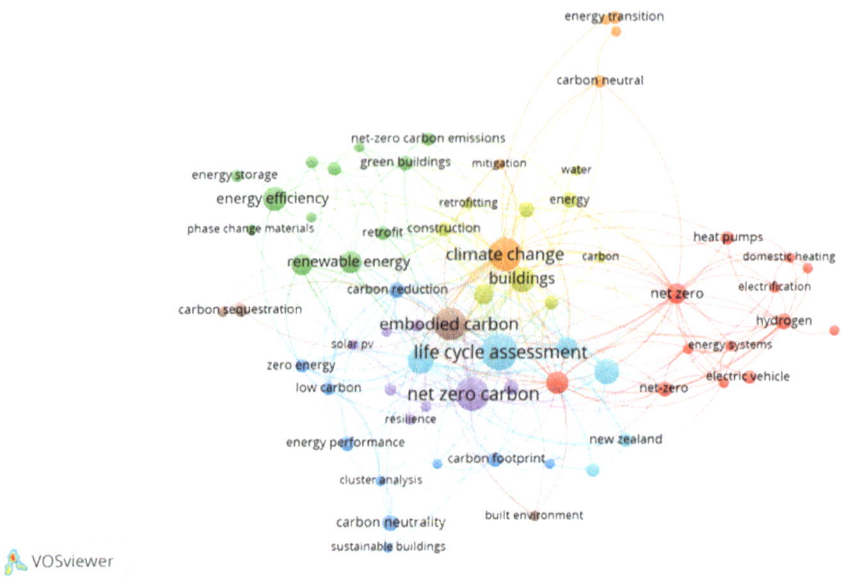

Fig. 4 Co-occurrence network of keywords

4.1.3 Country Scientific Production

The articles were produced by 1485 authors from 53 countries, as recognised by bibliometrix. The geographical distribution of papers, based on all author's affiliations is presented in Fig. 5. The co-occurrence of countries was mapped using all authors' affiliations using Biblioshiny software. Therefore, the number of frequencies is found to be greater than the number of articles selected for bibliometric analysis. In the figure, the grey colour indicates the non-availability of local authors, while the blue shades (from light blue to dark blue) indicate an increase in number of documents with local authors [26].

When considering the highly ranked countries, it is evident that the countries represent different continents such as America, Asia, Europe, and Australia. The articles were mainly produced by researchers from the UK with 416 frequencies, followed by China and the USA. Further, leading countries like Canada, Australia, and New Zealand have also shown a significant frequency in NZCB publications. This is not surprising as many scientific publications have identified that the same countries are leading the way in NZCB research [6, 27, 28]. Their efforts include a wide range of strategies, policy developments technological advancements, and practical implementations towards carbon neutrality by mid-century. The evolvement of these efforts and key global initiatives are discussed in the subsequent sections.

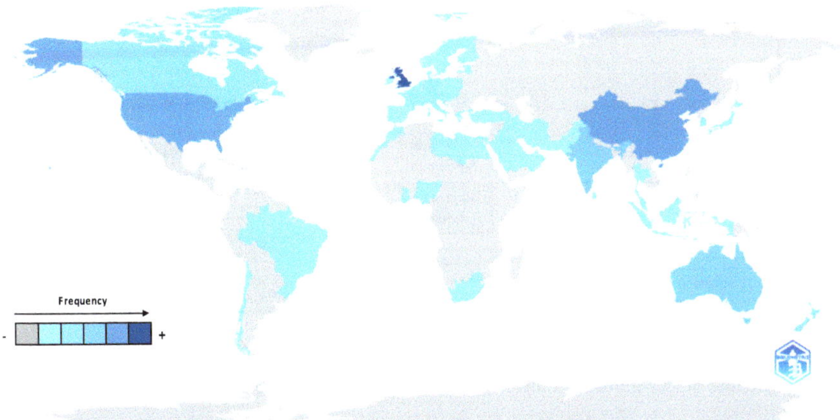

Rank	Country	Frequency	Rank	Country	Frequency
1	UK	416	6	New Zealand	55
2	China	187	7	Canada	46
3	USA	171	8	Germany	39
4	India	108	9	Italy	33
5	Australia	93	10	Malaysia	30

Fig. 5 Scientific production by country according to authors' affiliation

4.2 Notable Trends and Global Initiatives During the Evolution of Net Zero Carbon Worldwide

The evolution of global trends and key initiatives were identified through the content analysis of selected 34 papers and illustrated in Fig. 6. The first second and third analysis periods were based on 8, 11, and 15 number of publications, respectively. The trends were discussed under three timelines based on two key milestones; the Kyoto Protocol and the Paris Agreement. The Kyoto Protocol, which came into force in 2005, aimed at reducing GHG emissions to combat global warming and climate change. During the first commitment period of the Kyoto Protocol (2008–2012), industrialised countries were required to reduce their GHG emissions by an average of 5% compared to the 1990 level [29, 30]. However, most of the countries failed to lengthen their commitments and lost enthusiasm for the Protocol's continuation by 2011, at the end of the first commitment period, even though the Protocol was extended until 2020 [9]. The commitment to the Kyoto Protocol was further downturned when countries agreed to develop a new global climate agreement in 2012, later signed as the Paris Agreement in 2015 [9, 31]. The Paris Agreement was adopted by 196 countries at the Conference of Parties (COP) in 2015, and established targets to limit temperature increases below 2 °C or ideally 1.5 °C by 2050 [3]. Even though the Paris Agreement has not superseded the Kyoto Protocol, this chapter considered the climax period of the protocol (2005–2011) as the Kyoto Protocol era. The pre-Paris agreement (2012–2015) and post-Paris agreement (2015–2024) periods were

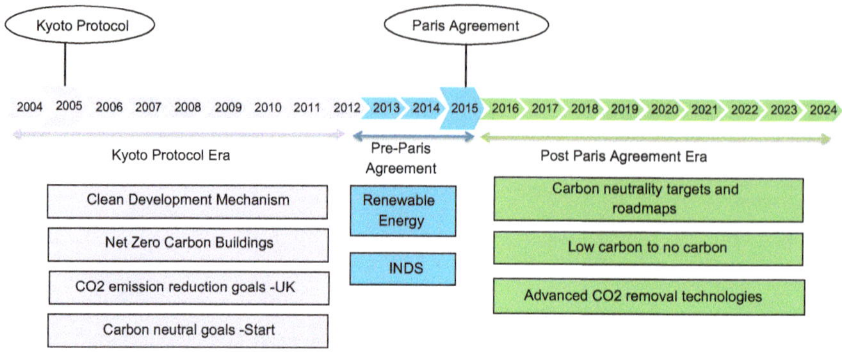

Fig. 6 Timeline of key global initiatives and carbon reduction efforts from 2004 to 2024

adopted subsequently in analysing the timeline and evolvement of net zero carbon in the construction industry.

4.2.1 The Kyoto Protocol Era (2005–2012)

People were aware of the importance of emission reduction by the time of the Kyoto Protocol. This has resulted from the continuous recognition and interest in climate issues from the establishment of the Intergovernmental Panel on Climate Change (IPCC) in 1988 [32]. The Kyoto Protocol was adopted in 1997 and entered into force in 2005, which marks the start of the first analysis period for the current study. However, the first appearance of the term net zero in the construction sector was evident in 2004 in a journal article which provided an overview of a commercially built carbon-neutral building [24]. Terms such as 'carbon neutrality' and 'self-sufficient energy' were noted from the publication which turned out to be common terms during the era. Once the Kyoto Protocol was enforced, the idea of carbon neutrality began to rise, with a focus on net zero energy buildings (NZEBs). Energy efficiency for new housing, renovations, and expansions was widely discussed in the construction sector by exploring pathways towards NZEBs [33–35].

The Kyoto Protocol also created the clean development mechanism (CDM), a market-based approach for carbon offsets [36]. Offsets emerged as a common mechanism to reduce carbon emissions at the firms and states and continued to link to the concept of carbon neutrality. Subsequently, nations and sub-nations started to adopt carbon-neutral goals, moving beyond the carbon offsets. In 2006, the UK set goals to reduce carbon emissions from its housing stock and introduced a policy to build zero carbon houses by 2016 [33]. In an attempt to promote environmentally sustainable homes in Canada, the Canadian government created an initiative of sustainable housing demonstration in 2006 [37]. This aimed to integrate advanced building technologies and practices to achieve near-net zero energy performance. Such initiatives during the Kyoto Protocol era started to make a shift from conceptual ideas to physical implementation. The immediate efforts began with carbon neutrality being associated with carbon offsets and the idea progressed towards more physical achievement of net zero energy/carbon buildings [9]. The works began to consider how to convert targets into policies and ways of implementing them.

4.2.2 Pre Paris Agreement Era (2012–2015)

The next period of the evolution begins after the first commitment period of the Kyoto Protocol. In 2009, it became obvious that the Kyoto Protocol was not politically tenable [9]. In 2011, countries started working on a new global climate agreement and the experts also believed that more stringent targets would offer several benefits [32]. Two key developments were noted during the period which shaped the transition towards net zero carbon in the built environment. First, countries and governments started discussing how to translate the temperature goals into a plan for reducing emissions [38]. Second, some countries began setting more stringent targets for carbon neutrality, ultimately for net zero carbon. As a result, Intended Nationally Determined Contributions (INDCs) were presented by countries to outline the country's planned efforts in reducing national emissions. These INDCs were converted into NDCs (shifted from intentions to commitments) when the Paris Agreement was signed in 2015.

The scholars still remained focused on energy systems and carbon reduction, according to the current analysis, while the governments were mapping out the net zero targets. Publications have continued to focus on NZEBs, but there has been a growing interest in the integration of renewable energy sources into buildings. Articles discussed different terms and definitions related to NZEBs, mainly focusing on energy production and consumption [39]. However, some researchers have expanded their research into NZCBs over NZEBs with technical, regulatory, and policy improvements being the main concerns [40, 41]. Overall, a notable disconnection between the literature and actual progress towards net zero carbon is evident during the pre-Paris Agreement period. Countries already started to set net zero carbon targets while the researchers remained focused on NZEBs with a narrowed attention towards net zero carbon.

4.2.3 Post-Paris Agreement Era (After 2015)

The Paris Agreement emerged as an explosion of net zero pledges. The literature on net zero carbon in the post-Paris Agreement period widely spread over different sectors of the economies and industries. In the built environment sector, more serious engagements in net zero carbon buildings have been observed in the post-Paris Agreement era. This increased attention to net zero carbon construction is also evident in the growing academic literature, with a noticeable rise in publications from 2015 to 2019 and a gradual increase thereafter, as shown in Fig. 3 based on bibliometric analysis.

Countries around the world have started presenting their roadmaps towards carbon neutrality by mid-century. Meantime, scholars began to engage more seriously with net zero carbon buildings in terms of definitions, technologies, policies, barriers, and enablers [42, 43]. Many researchers referred to the post-Paris period as a shift from a low-carbon era to a no-carbon era while some researchers have argued on the possibility of achieving carbon neutrality by 2050 [32, 44, 45]. Also, several studies on carbon removal technologies were carried out, providing a major focus on carbon estimation and associated reductions. Carbon estimation and low-carbon technologies have been considered over the life cycle of a building rather than limiting them to a specific stage. Moreover, energy-related literature related to low-carbon technologies such as energy efficiency and renewable energy has continued to dominate the field [46, 47]. However, the main focus was shifted towards net zero carbon emissions in the construction sector, especially after the Paris Agreement came into force.

5 Conclusion

This chapter aimed to establish a clear understanding of the concept of net zero carbon within the construction domain and provide a valuable historical context and evolution of carbon reduction efforts in the construction sector worldwide. The chapter introduced various concepts, terminologies, and definitions around the net zero carbon in construction industry and set a nuanced definition for subsequent discussions: "NZCBs is a state where the amount of carbon emissions associated with a building's embodied and operational impacts over the life of the building, including its disposal, are zero or negative".

The chapter conducted a systematic literature review to trace the evolution and global initiatives in the construction sector towards NZCBs and presented the findings of the bibliometric analysis and content analysis. The bibliometric analysis included 413 papers from academic literature in English and published by mid-July, 2024. The findings revealed that the publications on net zero carbon in the construction sector have been increasing since 2004, and gradually increased after 2019, resulting from the Paris Agreement. Scholars from the UK, China, and the USA have made a significant contribution to the literature, followed by India, Australia, and New Zealand. The content analysis is presented in a timeline where the Kyoto Protocol and Paris Agreement are identified as the key milestones of the evolution. The construction sector has transformed from low carbon to no carbon from the Kyoto Protocol era to post-Paris implementations. Carbon reduction efforts such as the clean development mechanism, NZEBs, and early stages of carbon-neutral goals from the Kyoto Protocol era have shifted to nationally determined contributions, roadmaps, and policies towards net zero carbon and advanced carbon removal technologies in the Paris Agreement era. A detailed discussion of these topics, including policies, technologies, and stakeholders in the net zero carbon construction industry is presented in the subsequent chapters of the book.

References

1. World Green Building Council (2019) Global status report for buildings and construction [Online]. World Green Building Council. Available https://worldgbc.org/article/2019-global-status-report-for-buildings-and-construction/. Accessed 17 Mar 2023
2. Tirelli D, Besana D (2023) Moving toward net zero carbon buildings to face global warming: a narrative review. Buildings 13:684
3. United Nations (2015) Paris Agreement [Online]. Available https://www.un.org/en/climatechange/paris-agreement. Accessed 17 Mar 2023
4. Ahmed AK, Ahmad MI, Yusup Y (2020) Issues, impacts, and mitigations of carbon dioxide emissions in the building sector. Sustainability 12:7427
5. Ganem-Karlen C, Barea-Paci GJ, Andreoni-Trentacoste SE (2023) What it takes to go net zero: why aren't we there yet? In: Sayigh A (ed) Towards net zero carbon emissions in the building industry. Springer International Publishing, Cham
6. Le A, Navodana R, Niluka D, Senaratne S (2023) Policy mapping for net-zero-carbon buildings: insights from leading countries. Buildings 13:2766
7. Þórólfsdóttir E, Árnadóttir Á, Heinonen J (2023) Net zero emission buildings: a review of academic literature and national roadmaps. Environ Res Infrastruct Sustain 3
8. Loveday J, Morrison GM, Martin DA (2022) Identifying knowledge and process gaps from a systematic literature review of net-zero definitions. Sustainability 14:3057
9. Green JF, Reyes RS (2023) The history of net zero: can we move from concepts to practice? Clim Policy 23:901–915
10. Hill S, Dalzell A, Allwood M (2020) Net zero carbon buildings: three steps to take now [Online]. ARUP. Available https://www.arup.com/perspectives/publications/research/section/net-zero-carbon-buildings-three-steps-to-take-now. Accessed 10 Apr 2024
11. Causone F, Tatti A, Alongi A (2021) From nearly zero energy to carbon-neutral: case study of a hospitality building. Appl Sci 11:10148
12. Green Building Council Australia (2021) A climate positive roadmap for the built environment
13. Whole Life Carbon Network (WLCN) (2021) Improving consistency in whole life carbon assessment and reporting: carbon definitions for the built environment. Buildings and Infrastructure, London
14. UK Green Building Council (UKGBC) (2019) Net zero carbon buildings: a framework definition, London, UK
15. Birkeland J (2003) 'Beyond Zero Waste', Societies for a sustainable future. In: Third UKM-UC international conference proceedings, Canberra, Australia
16. GBCA (2023) Carbon and climate definitions [Online]. Available https://gbca-web.s3.amazonaws.com/media/documents/carbon-and-climate-definitions.pdf. Accessed 10 Apr 2024
17. IGBC (2024) Net zero carbon definitions [Online]. Available https://www.igbc.ie/net-zero-carbon-definitions/. Accessed 10 Apr 2024
18. USGBC (2024) Net zero [Online]. Available https://www.usgbc.org/about/priorities/net-zero. Accessed 12 Apr 2024
19. GBCSA (2020) Getting to zero [Online]. Available https://gbcsa.org.za/wp-content/uploads/2020/08/Getting-to-Zero_2020.pdf. Accessed 10 Apr 2024
20. Davis J, Mengersen K, Bennett S, Mazerolle L (2014) Viewing systematic reviews and meta-analysis in social research through different lenses. SpringerPlus 3:511
21. Sohrabi C, Franchi T, Mathew G, Kerwan A, Nicola M, Griffin M, Agha M, Agha R (2021) PRISMA 2020 statement: what's new and the importance of reporting guidelines. Int J Surg 88:105918
22. Wafa W, Sharaai AH, Matthew NK, Ho SAJ, Akhundzada NA (2022) Organizational life cycle sustainability assessment (OLCSA) for a higher education institution as an organization: a systematic review and bibliometric analysis. Sustainability 14:2616
23. Senaratne S, Rodrigo N, Almeida LMMCE, Perera S, Jin X (2023) Systematic review on stakeholder collaboration for a circular built environment: current research trends, gaps and future directions. Resour Conserv Recycl Adv 19:200169

24. Jones D (2004) Leading by example: wind developer builds first commercial net zero carbon dioxide emissions building in the UK. Refocus 5:38–41
25. Fatma N, Haleem A (2023) Exploring the nexus of eco-innovation and sustainable development: a bibliometric review and analysis. Sustainability 15:12281
26. de Sousa Neto AR, de Carvalho ARB, Ferreira da Silva MD, Rêgo Neta MM, de Oliveira Sena IV, Almeida RN, Filha FSSC, Lima e Silva LL, da Costa GR, da Silva Lira IM, Portela DMMC, Oliveira e Silva AT, de Moura Rabêlo CB, da Costa Valle ARM, Moura MEB, de Freitas DRJ (2023) Bibliometric analysis of global scientific production on COVID-19 and vaccines. Int J Environ Res Public Health 20:4796
27. Elín Þ (2023) Net zero emission buildings a comprehensive review of pathways and strategies towards achieving net zero emissions. Environment and Natural Resources M.Sc., University of Iceland
28. Ohene E, Chan APC, Darko A (2022) Review of global research advances towards net-zero emissions buildings. Energy Build 266:112142
29. Aldy JE, Stavins RN (2013) 8 Designing the post-Kyoto climate regime. In: Joseph ES, Mary K (eds) The quest for security. Columbia University Press, New York, Chichester, West Sussex
30. Maamoun N (2019) The Kyoto protocol: empirical evidence of a hidden success. J Environ Econ Manag 95:227–256
31. UNFCCC (2012) Report of the conference of the parties on its seventeenth session, held in Durban from 28 Nov to 11 Dec 2011
32. Zhang Z, Hu G, Mu X, Kong L (2022) From low carbon to carbon neutrality: a bibliometric analysis of the status, evolution and development trend. J Environ Manag 322:116087
33. Banfill P, Peacock A (2007) Energy-efficient new housing—the UK reaches for sustainability. Build Res Inf 35:426–436
34. Beazant G (2006) The big green apple. Prof Eng 19:28–29
35. Hamilton IG, Summerfield AJ, Steadman JP, Stone A, Davies M (2010) Exploring energy integration between new and existing developments. Build Res Inf 38:593–609
36. Gillenwater M, Seres S (2011) The clean development mechanism: a review of the first international offset programme. Greenhouse Gas Meas Manag 1:179–203
37. Noguchi M, Athienitis A, Delisle V, Ayoub J, Berneche B (2008) Net zero energy homes of the future: a case study of the ÉcoTerraTM house in Canada
38. Matthews HD, Caldeira K (2008) Stabilizing climate requires near-zero emissions. Geophys Res Lett 35
39. Mlecnik E (2011) Defining nearly zero-energy housing in Belgium and the Netherlands. Energy Effi 5
40. Andrianopoulos E, Acha S, Shah N (2015) Achieving net zero carbon performance in a commercial building by aligning technical and policy alternatives—an UK case study
41. Berry S, Davidson K, Saman W (2014) Defining zero carbon and zero energy homes from a performance-based regulatory perspective. Energy Effi 7:303–322
42. Braune A, Geiselmann D, Oehler S, Ruiz Durán C (2019) Implementation of the DGNB framework for carbon neutral buildings and sites. IOP Conf Ser Earth Environ Sci 290:012040
43. Scherz M, Passer A, Kreiner H (2020) Challenges in the achievement of a net zero carbon built environment—a systemic approach to support the decision-aiding process in the design stage of buildings. IOP Conf Ser Earth Environ Sci 588:032034
44. Deutch J (2020) Is net zero carbon 2050 possible? Joule 4:2237–2240
45. Liu HY (2017) From low carbon to no carbon: building a sustainable dwelling through carbon neutrality. J Archit Plan Res 34:161–175
46. Kim D, Cho H, Mago PJ, Yoon J, Lee H (2021) Impact on renewable design requirements of net-zero carbon buildings under potential future climate scenarios. Climate 9:17
47. Lou H-L, Hsieh S-H (2024) Towards zero: a review on strategies in achieving net-zero-energy and net-zero-carbon buildings. Sustainability 16:4735

Net Zero Carbon Policies in the Construction Industry

N. Rodrigo and W. W. Jayathilaka

Abstract The buildings and construction industry was responsible for 30% of the total energy consumption and 26% of the global energy-related emissions that were related to energy and processes in 2022. Concerns about climate change are growing, and as a result, there is an increasing need to implement net zero carbon (NZC) regulations and policies in order to reduce the negative impact that construction activities have on the environment. To achieve a net zero impact on the environment, the policies of the NZC strive to achieve a balance between the amount of carbon dioxide (CO_2) that is released into the atmosphere and the amount that is removed or offset. Within the context of the construction industry, this chapter investigates the prospective outcomes, obstacles, and execution of NZC regulations and policies. The regulations and policies related to NZC in the construction industry of nine (9) countries named United States (US), Japan, Germany, South Korea, Canada, China, India, Indonesia, and Saudi Arabia were identified in this chapter. As well as challenges and opportunities related to regulatory compliance were identified by reviewing the current NZC regulations and policies, reports, previous research, and other related documents. Challenges in NZC policies are highlighted along with the opportunities. Action plan for carbon neutrality of the selected countries has been mapped within this chapter. The findings underline that how the developed and developing countries are reflected in their actions in the carbon neutrality path. This study provides a foundation for achieving carbon neutrality of the top carbon-emitting countries through NZC regulations and policies compliance and development. Further, the discussion extends to the need for collaboration between the government, policymakers, construction industry practitioners, and researchers to overcome challenges when stepping forward in the carbon neutralisation path.

Keywords Net zero carbon · Regulations and policies · Challenges · Opportunities

N. Rodrigo (✉) · W. W. Jayathilaka
School of Architecture and Civil Engineering, The University of Adelaide, Adelaide, Australia
e-mail: navodana.rodrigo@adelaide.edu.au

N. Domingo et al. (eds.), *Global Net Zero Carbon Practices in Construction*, Green Energy and Technology, https://doi.org/10.1007/978-981-96-3853-6_3

1 Introduction

The term "zero energy building" (ZEB) refers to a residential or commercial structure that has significantly decreased its energy requirements by implementing efficiency improvements, to the extent that renewable energy sources may meet the remaining energy requirements [1]. In addition to being extremely resource-efficient, Net Zero Energy Buildings (NZEB) balance all of the direct and indirect carbon emissions that are caused throughout its construction and operation process [2]. Through the facilitation of the accomplishment of NZEB, renewable energy sources can contribute to the decarbonisation of the economy [1].

Numerous parties such as policymakers, general public, and investors would explicitly acknowledge carbon as a greenhouse gas (GHG) that has major negative externality [3]. Moreover, the majority of GHGs that are produced in the construction industry are caused by carbon dioxide (CO_2) emissions that occur during the construction and operations of buildings [2]. According to Ministry of Environment, Forest and Climate Change Government of India (MoEFCC) [4], around 39% of all yearly emissions from energy sources are produced by the construction and operation of buildings, which consumes 36% of all global energy. About 145 countries, accounting for nearly 90% of the world's emissions, have declared or are contemplating net-zero targets as of November 2023 [5]. According to the climate action tracker website, China, the European Union, the United States (US), and India are the responsible countries for more than half of the world's greenhouse gas emissions. According to Bhattacharya et al. [6], "net-zero" emission goes beyond zero emission, and by specifying future target years, many countries have proposed net-zero emission targets through legislation.

Within the realm of promoting energy efficiency in buildings, the standards and policies are essential instruments [7]. In order to limit carbon emissions, a country requires a transparent plan to reach net-zero emissions and an effective monitoring system [8]. Standards and regulations for building energy efficiency are examples of policy measures that are frequently utilised to limit the amount of energy that is consumed in buildings [7]. Regulations and policies should be devised to increase energy efficiency in buildings in order to promote NZEB and this will ultimately serve for carbon neutrality. Therefore, this chapter is designed to explore the complex landscape of NZC regulations and policies as well as challenges and opportunities related to regulatory compliance, offering a comparative analysis between developed and developing countries. By examining the regulatory frameworks, it sheds light on the diverse approaches these nations employ to address carbon emissions within the construction sector.

Table 1 Country selection for the review

Economy	Country	World Resource Institute 2021	EDGAR 2021	Worldometer 2022
Developing	China	1	1	1
Developed	US	2	2	2
Developing	India	3	3	3
Developed	Japan	5	5	5
Developing	Indonesia	7	6	6
Developed	Germany	8	8	8
Developed	South Korea	10	9	9
Developing	Saudi Arabia	9	10	10
Developed	Canada	11	11	11

2 Methodology

Document review was employed as the methodology for this chapter with the aim of mapping carbon reduction regulations and policies in developed countries and developing countries. Document review has been recognised as an effective research technique that has been used by many researchers for a long period of time. With this methodology, a variety of materials are analysed, such as books, conference papers, journal articles, and institutional reports [9].

This chapter was built in three steps. As the first step, developed and developing countries have been identified with the support of the United Nations World Economic Situation and Prospects [10]. Simultaneously, developing countries and developed countries that are most responsible for world's greenhouse gas emissions have been identified using data given by the World Resource Institute (2021), Emissions Database for Global Atmospheric Research—EDGAR (2021), and Worldometer (2022). Subsequently, among the top eleven (11) countries of the list, five countries that have developed economies and four countries that have developing economies were incorporated into the review (Table 1).

As the last step, carbon reduction regulations and policies in selected countries along with the challenges and opportunities in regulatory compliance have been discussed. Based on the findings of the document review, action plans for carbon neutrality in each country were presented.

3 Findings and Discussion

Decisive actions to reduce the emissions of GHG, mainly CO_2, will help to prevent the most extreme effects of climate changes. Therefore, a significant number of countries have already taken the required steps to attain carbon neutrality in future. This topic presents the findings, and the discussion related to carbon reduction regulations and policies, net-zero targets that have been set by selected developed countries and developing countries as well as the challenges and opportunities in regulatory compliance.

3.1 Carbon Reduction Regulations and Policies in Developed Countries

This section presents the carbon reduction regulations and policies in the US, Japan, Germany, South Korea, and Canada, which have developed economies along with the action plan for carbon neutrality of each country.

US

The US government has established novel goals to tackle the climate crisis. For example, (a) to reduce US GHG emissions 50–52% below 2005 levels by 2023, (b) to achieve 100% carbon pollution-free electricity by 2035, and (c) to achieve net-zero emissions by 2050 [11]. In order to achieve these goals, several steps have been taken including establishment of Inflation Reduction Act 2022, Bipartisan Infrastructure Law (known as Infrastructure Investment and Jobs Act 2021 now), investments on clean energy manufacturing and clean power projects, Affordable Home Energy Earthshot, updated energy efficiency standards for appliances and equipment. United States Environmental Protection Agency (EPA) has introduced Greenhouse Gas Reporting Program (GHGRP) in 2009 where the large GHG mission sources, fuel and industrial gas suppliers, and CO_2 injection sites (approximately 8000 facilities) in the US are required to report their GHG data and these are made available to the public in annually [12].

Energy Policy Act was firstly introduced in 2005, which discussed ways to reduce GHG emissions and focuses on energy reductions for products and buildings [13]. However, this act had not included any indication of targets to achieve net-zero [14]. The Long-Term Strategy of the US introduces pathways to achieving net-zero GHG emissions by 2050 [15]. It includes five complementary technological transformations including (1) decarbonising electricity, (2) switching to clean fuels like hydrogen and sustainable biofuels, (3) reducing energy waste, (4) reducing non-CO_2 emissions such as methane, hydrofluorocarbon, nitrous oxide, and others, and (5) scaling up CO_2 removal through lad carbon sinks and engineered strategies [16]. Figure 1 presents the US's action plan for carbon neutrality by 2050.

US is a country having 50 states, each state maintains a separate climate action plan. Out of these, 33 states have already released a climate action plan that includes GHG emissions reduction targets as well as detailed actions the state can take to meet those goals [17]. California is one of the states that have been in the forefront by legislating the Senate Bill 32—California Global Warming Solutions Act of 2006 to reduce emissions 40% below 1990 levels by 2030, to achieve carbon neutrality as soon as possible and no later than 2045, then achieve and maintain net-negative

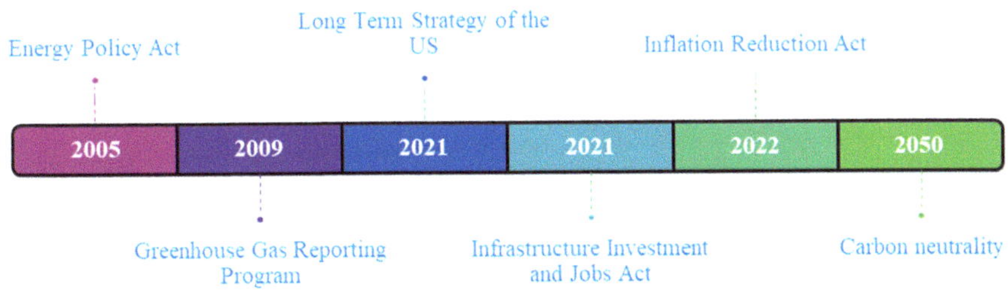

Fig. 1 Action plan for carbon neutrality in 2050 by US

emissions thereafter [18]. Regional Greenhouse Gas Initiative, a cap-and-trade agreement, has been signed among 11 Northeast states (Connecticut, Delaware, Maine, Maryland, Massachusetts, New Hampshire, New Jersey, New York, Rhode Island, Vermont, and Virginia), which caps climate-damaging emissions from power plants [19].

Japan

Japan is aiming at reducing its GHG emissions 46% below 2013 levels by 2030 and will continue their strenuous efforts in meeting the ultimate goal of achieving net-zero by 2050 [20]. Japan has already achieved a reduction of approximately 20%, and it is believed that Japan's reduction is on track [21]. According to Kuwabara et al. [22], Japan emitted a net of 1230 $MtCO_2e$ in 2017, most of which had originated from five sectors: power (37%), industry (36%), transportation (17%), buildings (10%), and agriculture (4%) (where the total will be 100% when negative emissions from land use, land-use change, and forestry are taken into consideration).

Based on Japan's Long-Term Strategy under the Paris Agreement (after their cabinet decision in 2019) and the Integrated Innovation Strategy 2019, Environment Innovation Strategy was introduced in January 2020 to create innovations in the fields of energy and environment [23]. This provided innovating action plans for five prioritised areas covering energy transformation, transportation, industry, business/household, and agriculture. It included mechanisms such as ensuring renewable energy as the main power source, low-cost hydrogen supply chain, CO_2 capture and carbon recycling, green mobility modalities, energy sharing mechanisms among others. In June 2020, Japan introduced the Green Growth Strategy, which is an industrial policy towards achieving a positive cycle of economic growth and environmental protection [24]. It includes five key policy tools related to grant funding, tax incentives, guidance policy on finance, regulatory reform, and international collaboration. As a result, Green Innovation Fund was introduced with 2 trillion yen to encourage companies to take a greener approach in how they operate [25]. Green Innovation Fund has already contributed to several Research and Development (R&D) projects, for example, wind power generation, solar cell development, large-scale hydrogen supply chain, fuel ammonia supply chain, smart mobility, CO_2 reduction/absorption for agriculture/forestry, carbon recycling, and so forth [26].

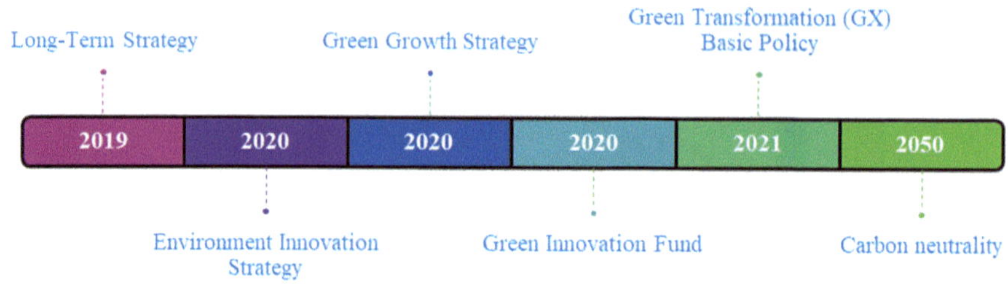

Fig. 2 Action plan for carbon neutrality in 2050 by Japan

In 2022, Japan introduced Green Transformation (GX) Basic Policy, which is an investment roadmap for 150 trillion yen of public–private financing over the next ten years to transform 22 industrial sectors to meet carbon neutrality. Japan is also introducing a carbon pricing scheme, launched in April 2023, where an emissions trading system (ETS) was introduced in 2023 on a voluntary basis, followed by full-scale operation from around 2026/2027 [27]. The action plan of Japan towards the pathway to carbon neutrality by 2050 is depicted in Fig. 2.

Germany

Germany's intention is to reach net GHG neutrality by 2045 and to achieve that they have set interim targets of reducing emissions by 65% by 2030 and 88% by 2040 compared to 1990 levels [28]. Germany is aiming for achieving net-negative emissions post-2050. Germany is looking forward to presenting a strategy towards end of 2024 on setting a target to achieve by 2060 related to net-negative emissions, for example, introducing carbon sinks that can capture CO_2 from the atmosphere and store underground permanently [29].

Germany established a Renewable Energy Act in 2000 to encourage renewable energy use. As a result, 52% of the country's power consumption was covered by renewables by 2023 [28]. Germany is well aligned to bringing the renewables share in power consumption to 80% by 2030. Climate Action Programme 2030 was adopted on 9 October 2019 by the Cabinet of Germany. It focuses on introducing CO_2 pricing (national CO_2 emissions trading regulated by Fuel Emissions Trading Act), providing financial relief for renewable energy contributors, providing tax relief for energy-efficient renovations, promoting electric mobility, and encouraging lower CO_2 emissions from production in various industries such as steel (Steel Action Concept) [30].

The Climate Change Act 2021 (Bundes-Klimaschutzgesetz) has been introduced to tighten climate regulations and enshrine in law to achieve GHG neutrality by 2045 [31]. The act has established new milestones to increase annual reduction targets per sector from 2023 to 2030, enshrine annual reduction targets for 2031–2040 into law, determine annual reduction targets by sector for the period 2031–2040 by 2024, and determine annual reduction targets for the period 2041–2045 by 2032 [32]. Germany has set up Climate and Transformation Fund of 36 billion euros to finance climate and energy transition policy measures [33]. In addition, EUR 8 billion has been allocated

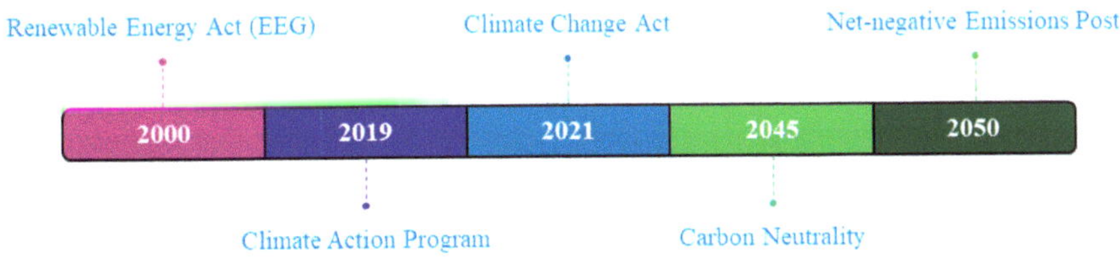

Fig. 3 Action plan for carbon neutrality in 2045 by Germany

for climate protection measures in all sectors between 2022 and 2025 [32]. Germany's action plan towards carbon neutrality by 2045 and beyond that is presented in Fig. 3.

South Korea

According to IEA [32], the building industry was responsible for 8% of the South Korean's CO_2 emissions from fuel burning in the year 2019. As stated in the climate action tracker, all new buildings in South Korea are required to satisfy zero energy construction requirements by 2025, and Korea is progressively implementing more stringent energy-saving designs. By 2030, the goal is to reduce carbon emissions from buildings by 35.8 $MtCO_2e$ compared to the baseline in 2016 [5].

As reported in International Energy Agency website, in 2001, South Korea introduced Building Energy Efficiency Certification to boost energy efficiency, by valuating and certifying buildings based on their CO_2 emissions and energy use [34]. In 2003, Building Energy Code Compliance was initiated by the South Korean government. Building GHG and Energy Target Management Scheme was established in 2010, by the government collaboratively with the industry to reduce energy consumption and greenhouse gas emissions by 30% compared to 2020 and meet the national target. The Green Building Construction Support Act was set up in 2012 to promote green growth with low-carbon emissions and improve people's wellbeing by reducing building GHG emissions and boosting green buildings construction [35].

South Korea is one of the countries that have committed to become carbon neutral by the year 2050, introducing a Green New Deal in year 2020. The announcement of the Green New Deal was made by the government in July 2020 to speed up Korea's transition to a more sustainable energy system [36]. The Framework Act on Carbon Neutrality and Green Growth (Carbon Neutrality Act) was enacted in August 2021 to enshrine South Korea's goal of achieving carbon neutrality [5]. As stated by International Energy Agency [37], building energy management systems (BEMS) are also expected to be mandated beginning in 2025 with the goal of achieving net zero energy use in buildings, industries, and cities. The action plan towards carbon neutrality of South Korea by 2050 is presented in Fig. 4.

Canada

Canada has two complimentary climate targets, (1) to achieve net-zero GHG emissions by 2050 and (2) to reduce national GHG emissions by 40–45% below 2005 levels by 2030 [12].

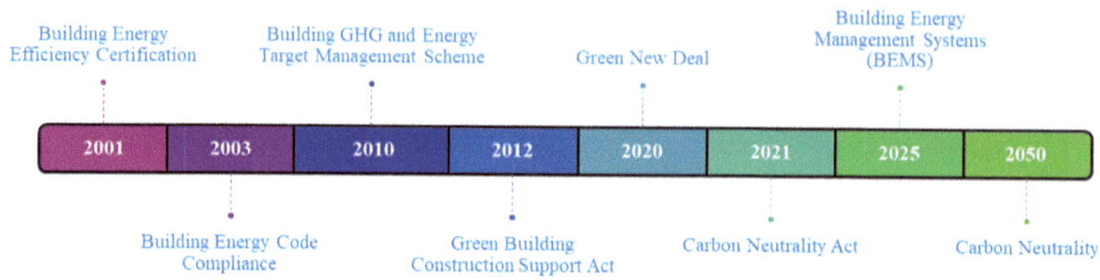

Fig. 4 Action plan for carbon neutrality in 2050 by South Korea

Building on the actions in Canada's strengthened climate plan published in 2020 and the Pan-Canadian Framework introduced in 2016, the 2030 Emissions Reduction Plan was introduced in 2022 that reflects how the Paris Agreement target to reduce emissions by 40–45% from 2005 levels by 2030 can be achieved [14, 38]. The 2030 Emissions Reduction Plan is an ambitious and achievable roadmap that outlines a sector-by-sector path for Canada to achieve GHG emissions target. According to the Progress Report published in 2023 on the 2030 Emissions Reduction Plan, Canada is on track to exceed the previous climate target of 30% below 2005 levels by 2030.

A discussion paper on the Canada Green Buildings Strategy highlights how the building materials and construction practices can be improved to provide a low-carbon, climate-resilient, and energy-efficient building sector (Natural Resources Canada 2024). As a result, $800 million has been introduced under Canada Greener Homes Affordability Program, which replaced Canada Greener Homes Grant to help low-to-median-income Canadians to upgrade their homes to reduce their energy bills [39].

Canadian Net-Zero Emissions Accountability Act was implemented in June 2021 [40]. The main purpose of this Act is to promote transparency, accountability, and immediate and ambitious action in achieving Canada's net-zero GHG emissions reduction targets by 2050. In order to accomplish this, the act provides a framework that includes planning, progress reporting, and third-party evaluations informed by independent expert advice [41]. Figure 5 summarises Canada's action plan for carbon neutrality.

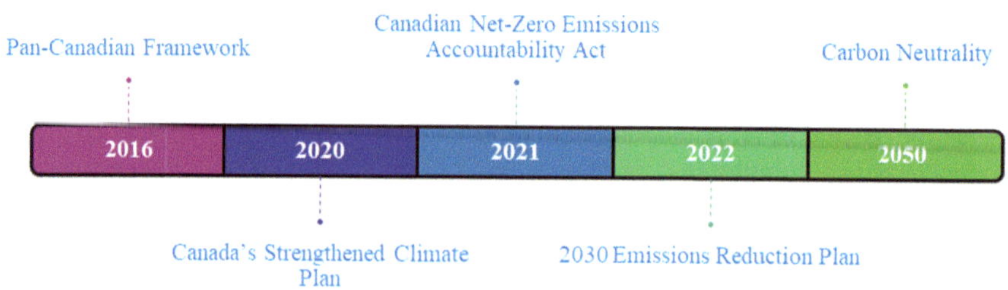

Fig. 5 Action plan for carbon neutrality in 2050 by Canada

3.2 Carbon Reduction Regulations and Policies in Developing Countries

This section presents the carbon reduction regulations and policies in China, India, Indonesia, and Saudi Arabia which have developing economies along with the action plan for carbon neutrality of each country.

China

Achieving global net-zero emissions through the transition to sustainable power generation is considered as a shared responsibility among countries like China and India, which have major populations [42]. As stated by Feng and Berkeley [43], the formulation of policies should be coordinated in order to achieve zero net energy development in China. Moreover, the authors stated that China's building sector currently accounts for almost 20% of its total primary energy consumption, and this proportion is projected to increase further in future as a result of rapid urbanisation [43]. To achieve carbon neutrality by the year 2060, China had focused its efforts already and in 2016 China has taken actions to enforce Urban Climate Change Action Plan, and green development was given a significant amount of attention through this plan [44]. In 2019, the Technical Standard for Nearly Zero Energy Buildings GB/T51350-2019 was issued by the Ministry of Housing and Urban–Rural Development (MOHURD) of China which defined the concepts related to ZEB, and in the same year, the principles of energy-capable buildings were introduced in TB/CABEE 003–2019, which was issued [45].

China's first zero-carbon building standard was published in 2021 by the Tianjin Low-Carbon Development Research Centre: the Zero Carbon Building Identification and Evaluation Guidelines. New guidelines for the total lifetime of zero-carbon buildings were introduced in July 2023 by the Chinese Ministry of Housing and Urban–Rural Development. These standards cover design, construction, operation, and assessment of zero-carbon buildings [46]. As a result of these persistent advancements, NZC building determination is becoming easier, and zero-carbon development ideas are being incorporated into regional planning and development in China. The action plan of China towards the pathway to carbon neutrality by 2060 is depicted in Fig. 6.

India

Achieving the worldwide net-zero aim will need the energy transition to be spearheaded by developing Asian countries like India and China in the next decades [6]. Over the last 20 years, the Indian government has launched a number of new initiatives and laws that will help pave the way for construction industry to achieve net-zero carbon emissions [2]. Moreover, the government has already established several regulations and policies to encourage the use of renewable energy sources [8].

In 2001, India established the Energy Conservation Act, which makes provisions for the conservation of energy and the efficient use of energy. The energy conservation

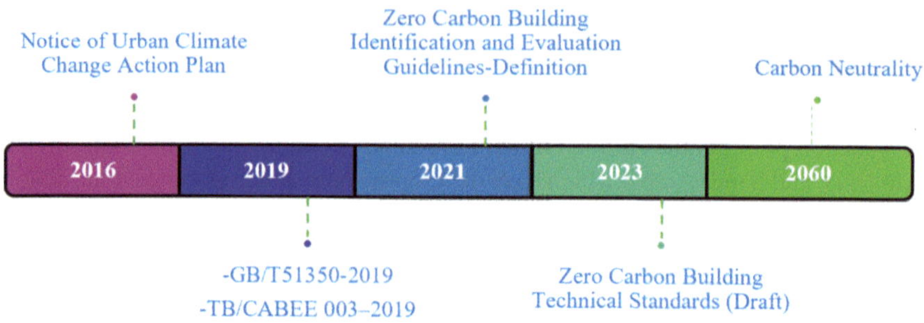

Fig. 6 Action plan for carbon neutrality in 2060 by China

building code (ECBC) was introduced in 2007 to address the energy challenges that were prevalent throughout India by Bureau of Energy Efficiency (BEE). In 2008, National Action Plan on Climate Change (2008) was introduced and it led to promoting low-carbon development within the country. The Indian government has launched National Solar Mission (2010) and Swachh Bharat Mission (2014) to help pave the way towards NZEBs. On 25 June 2015, Smart City Mission was launched by the government of India to support urban areas to provide essential infrastructure, a healthy environment, and adequate housing to the people and further updates to ECBC were made in 2017 by the BEE India. Intending to reach zero energy building status, the IGBC published the Net Zero Energy Buildings Rating System in 2018. In comparison to the national baseline, this rating system proposes measures to minimise yearly energy usage in existing buildings by 25–30% while simultaneously reducing costs by 30% [47]. In the same year, a new residential energy conservation building code named Eco Niwas Samhita-Part I was introduced that allowed the ECBC code to be extended for residential developments [48]. In addition, another two policies named the National Energy Policy and the Sustainable Public Procurement were established in the same year by the Indian Government.

The BEE also introduced the energy efficiency label for residential projects in 2019 which established the minimal criteria for energy efficiency [49]. In the same year, the BEE introduced Eco Niwas Samhita-Part II in order to ensure that residential buildings comply with the code for electro-mechanical and renewable energy systems. Moreover, IGBC introduced the Guidance Framework for NZCB in 2019 and it defined NZCB as extremely energy-efficient buildings that neutralise all carbon emissions, both directly and indirectly, caused by their construction and operation [46]. With the Energy Conservation Act 2022, the scope of the construction sector and the building regulations has been widened. By enforcing several regulations and policies, the Indian government has pledged to attain carbon neutrality as soon as possible and no later than by the year 2070 [50]. India's action plan towards the pathway to carbon neutrality by 2070 is presented in Fig. 7.

Indonesia

Retraining and upskilling policies could enable employees to fully benefit from new job opportunities in a low-carbon economy, particularly in the construction

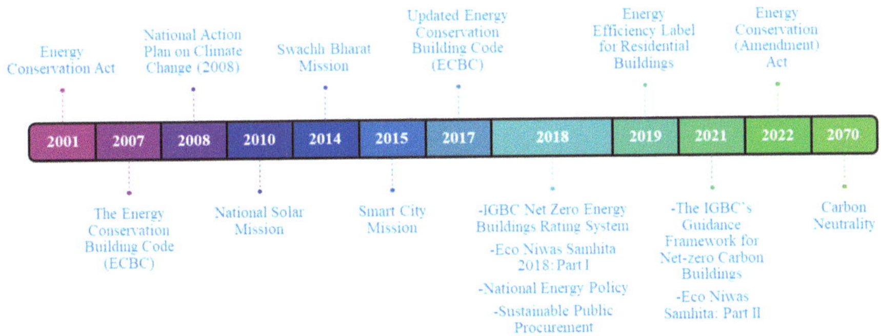

Fig. 7 Action plan for carbon neutrality in 2070 by India

industry which may experience a large increase in employment [51]. According to Svendsen [52], in Indonesia, 4% of all carbon emissions from energy sources are from the construction industry. Indonesia needs better rules to ensure that government policies are aligned and implemented effectively to meet its carbon neutrality targets [53]. To guide Indonesia's building and construction industry towards greater energy efficiency, strong policies are needed to overcome any challenges and encourage the adoption of new technologies [52]. Moreover, Indonesia needs better rules to ensure that government policies are aligned and implemented effectively to meet its carbon neutrality targets [53].

The Indonesian government enforced a policy named Energy Mix Policy in 2006 for the construction industry in Indonesia [44]. Then to increase energy efficiency in the building sector, Indonesia implemented several regulations in 2009 (Regulation no. 70), 2012 (Minister of Energy and Mineral Resources Regulation no. 14) and 2014 (Minister of Energy and Mincral Resources Regulation no. 18). In 2014, Indonesia introduced a National Energy Policy to achieve energy independence and resilience [54]. Presidential Regulation no. 22 (General National Energy Plan) and Minister of Energy and Mineral Resources Regulation no. 57 were issued in 2017. As stated by Wahyuni and Ardiansyah [54], the establishment of goals and objectives policies as starting guides for the General National Energy Plan was the objective of the National Energy Policy. In 2016, another regulation to increase energy efficiency in the construction sector Governmental Regulation no. 16 was established [52].

Indonesia's carbon emissions reduction strategy set up a target of a 29% reduction in Indonesia's carbon emissions by the year 2030. This must be accomplished for Indonesia to meet its carbon output goal of reaching net-zero emissions by the year 2060 [53]. Figure 8 presents Indonesia's action plan for carbon neutrality by 2060.

Saudi Arabia

Saudi Arabia is striving to achieve net zero emissions by 2060 [55]. They have pledged to reduce its CO_2 emissions by 278 million tonnes per year by 2030. According to Kamboj et al. [56], Saudi Arabia is aiming at having 50% of its electricity generated from renewable sources by 2030 indicating the country's initiatives to have a greener and more sustainable landscape.

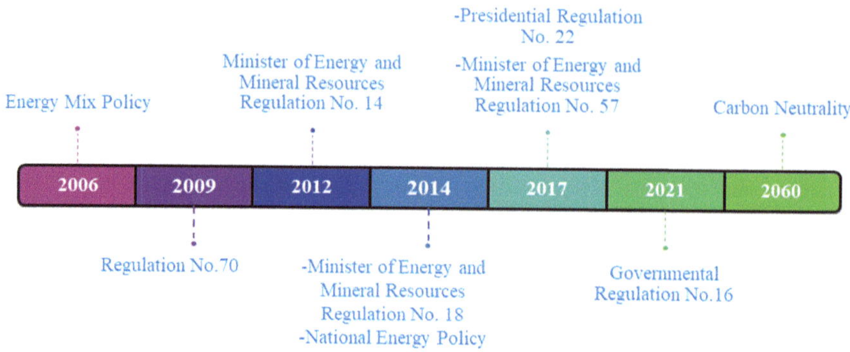

Fig. 8 Action plan for carbon neutrality in 2060 by Indonesia

Saudi Energy Efficiency Center (SEEC) was founded in 2010 to rationalise and increase energy efficiency in production and consumption while preserving Kingdom of Saudi Arabia's natural resources. It mainly involved (1) introducing new regulations (licencing, legal framework, incentives) and standards, (2) investing in human capital for building capabilities in the energy sector (certifications, training), and (3) promoting public awareness campaigns to enhance energy consumption efficiency [57]. SEEC has mandated thermal insulation in new buildings and increased minimum energy performance standards for building appliances (e.g., air conditioners) [58].

In 2017, National Renewable Energy Program (NREP) was launched under Vision 2030 [59], to include 50% renewable technologies and 50% natural gas in the power system by 2030. Saudi Arabia is constructing a carbon capture and storage (CCS) hub in Jubail to be launched in 2027 to capture 9 million tonnes of CO_2 annually with the target of capturing and storing 44 million tonnes by 2035 [58]. Circular Carbon Economy (CCE) National Program was launched in 2020 at the G20 Saudi Arabia Summit 2020. Its key strategic objectives are (1) climate protection, (2) socio-economic impact, and (3) global leadership. CCE focuses on activities related to developing roadmaps for key CCE technologies, supporting domestic industries in reducing emissions, promoting CCE framework among others [60]. Figure 9 shows the action plan for Saudi Arabia to reach carbon neutrality by 2060.

3.3 *Challenges and Opportunities in Regulatory Compliance*

In the route for carbon neutrality, there are numerous opportunities in different aspects, but most of the countries are also facing many challenges as well. In this section, challenges and opportunities in regulatory compliance related to net-zero carbon in the construction industry have been discussed.

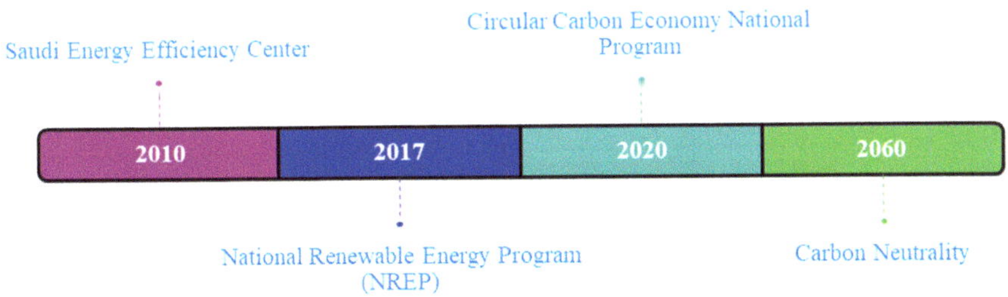

Fig. 9 Action plan for carbon neutrality in 2060 by Saudi Arabia

Most countries joined forces in December 2015 under the Paris Agreement to address climate change and it specified that each country brings the greatest amount of effort to achieve carbon neutrality globally [61]. A country's climate actions and policies are considered "1.5 °C Paris Agreement compatible" when they align with the Paris Agreement's temperature limit of 1.5 °C. According to Climate Action Tracker [5], Table 2 shows "1.5 °C Paris Agreement compatible" of selected countries which is the 1.5 °C target set by the Paris Agreement.

According to the Climate Action Tracker website, there are not many countries that meet 1.5 °C Paris Agreement compatible. When considering Table 2, even the most responsible countries for world's GHG emissions currently do not have sufficient policies and actions to achieve net-zero targets. For example, China falls in "highly insufficient" overall rating category, which means the country should be more concerned about possible ways to achieve 1.5 °C target set by the Paris Agreement.

In order to achieve carbon neutrality by 2045, Germany needs to invest in physical assets. According to Engel et al. [62], the total investments in physical assets required for the transition to carbon neutrality include additional investments of €1 trillion and a redirection of investments of €5 trillion. The investments will be used to replace ageing infrastructure, equipment, and buildings. Despite its efforts to achieve carbon neutrality and energy security, India is also encountering a number of challenges [63]. India, which has the greatest potential for promoting zero-carbon energy, is one of the many countries that still do not have access to inexpensive financing. The implementation of low-carbon solutions is hampered by the lack of low-cost financing availability. In order to overcome this obstacle, it is necessary to implement novel financial risk transfer mechanisms [3]. Policy in Indonesia does not reflect the country's stated goal of achieving net-zero carbon emissions. Moreover, a shortage of finance and demand leakage to other countries that have significant tracts of rainforests, impacts the progress of emissions to fluctuate considerably [51]. To achieve net-zero carbon emissions, renewable and carbon-free energy sources ought to be encouraged by the government [54]. Countries that have less developed economies also have the lowest global greenhouse gas emissions, but they also suffer severe development obstacles that make it difficult for them to participate in prospects for clean energy without access to affordable financing [64].

Table 2 1.5 °C Paris Agreement compatible

Country	Policies and action	Climate finance	Overall rating	Net zero target
China				2060
US				2050
India				2070
Japan				2050
Indonesia				2060
Germany				2045
South Korea				2050
Saudi Arabia				2060
Canada				2050

Legend of the rating	
1.5 °C Paris Agreement compatible	
Almost sufficient	
Insufficient	
Highly insufficient	
Critically insufficient	
Not assessed	
Not Applicable	

Source Adapted from Carbon Tracker, 2024

When considering the opportunities, as a result of the shift to the reduction of carbon emissions all over the world, the building sector also became more receptive to the idea of transitioning to a zero-carbon pathway [36]. Japan is planning towards contributing to lending approximately USD 9 billion related to climate finance through the provision of credit enhancements to the World Bank and the Asian Development Bank [21]. Germany is set to have net-negative GHG emissions post-2050, which would involve CO_2 removal from the atmosphere to help reach international climate targets. One of the key focus is on developing technical carbon sinks by 2035, 2040, and 2045, which would support natural sinks such as forests or peatland in Germany [29]. In Indonesia, energy efficiency investments support effectively at reducing building emissions at relatively low costs in the short term [51]. Moreover, many countries have advocated for the development of mandated zero emissions buildings by the year 2030, in conjunction with the implementation of energy efficiency programs that are intended to minimise emissions of GHGs [65].

4 Conclusion and Further Research

This chapter aims to provide insight into carbon reduction regulations and policies in developed and developing countries as well as challenges and opportunities in regulatory compliance in these countries. A document review was employed to find applicable details for the above content. Nine countries that have developing economies and developed economies were selected for the review. In addition, data availability in English was considered as a main selection criterion when selecting countries. Documents related to net-zero carbon regulations and policies in countries named US, Japan, Germany, South Korea, Canada, China, India, Indonesia, and Saudi Arabia were reviewed in this study.

As of November 2023, approximately 145 countries had either announced or were in the process of considering net zero targets, which comprise nearly 90% of global emissions [5]. According to researchers, nearly 40% of the total energy usage in the world is attributed to buildings. In addition to being extremely resource-efficient, a net-zero building is also capable of neutralising all of the direct and indirect carbon emissions that are caused by its construction and operation. Regarding the entire amount of activities that governments have taken up to this point, there is still a significant gap between what they pledged to do and what they have actually executed. The carbon neutrality in each country depends on climate action in other countries. Therefore, countries should pay more attention to the policies pertaining to energy efficiency, which account for the majority of savings in Net Zero Energy Buildings. However, opportunity and trade-off identification stemming from decarbonisation should be the primary emphasis, rather than merely determining the least-cost policy implementation. In order for vulnerable populations to be able to deal with rising living expenses and take advantage of the economic opportunities that are emerging in the low-carbon economy, the government must implement social programmes that provide income support and reskilling opportunities.

References

1. Madathil D, Ravikumar Pandi V, Nair MG, Jamasb T, Thakur T (2021) Consumer-focused solar-grid net zero energy buildings: a multi-objective weighted sum optimization and application for India. Sustain Prod Consumption 27:2101–2111. https://doi.org/10.1016/j.spc.2021.05.012
2. Indian Green Building Council (IGBC) (2021) IGBC'S guidance framework for net zero carbon buildings (issue Nov). https://www.ukgbc.org/ukgbc-work/net-zero-carbon-buildings-framework/
3. Chaturvedi V (2021) A vision for a net-zero energy system for India. Energy Clim Change 2(Dec 2020):100056. https://doi.org/10.1016/j.egycc.2021.100056

4. Ministry of Ministry of Environment Forest and Climate Change Government of India (MoEFCC) (2022) India's long-term low-carbon development strategy
5. New Climate Institute (2024) Net zero target evaluations. Climate Action Tracker. https://climateactiontracker.org/global/cat-net-zero-target-evaluations/
6. Bhattacharya S, Banerjee R, Ramadesigan V, Liebman A, Dargaville R (2024) Bending the emission curve—the role of renewables and nuclear power in achieving a net-zero power system in India. Renew Sustain Energy Rev 189(Nov). https://doi.org/10.1016/j.rser.2023.113954
7. Yu Z, Xu W, Chen X, Sun D, Zhang J, Lu F, Liu C (2019) Progress in energy efficiency standards of residential buildings in China's severe cold and cold zones. E3S Web Conf 111(2019). https://doi.org/10.1051/e3sconf/201911103032
8. Kumar S, Regassa Jufar S, Kumar S, Foroozesh J, Kumari S, Bera A (2023) Underground hydrogen storage and its roadmap and feasibility in India toward net-zero target for global decarbonization. Fuel 350(Jan):128849. https://doi.org/10.1016/j.fuel.2023.128849
9. Morgan H (2022) Conducting a qualitative document analysis. Qual Rep 27(1):64–77. https://doi.org/10.46743/2160-3715/2022.5044
10. United Nation (2024) United Nations World economic situation and prospects. In: United Nation Department of Economic and Social Affairs (issue). https://www.un.org/development/desa/dpad/publication/world-economic-situation-and-prospects-2024/
11. The White House (2021) Whitehouse.pdf. President Biden's historic climate agenda. https://www.whitehouse.gov/climate/
12. Energy Policy Act (2024) Green gas reporting program (GHGRP). https://www.epa.gov/ghgreporting
13. Energy Policy Act (2005) Energy policy act of 2005. https://www.congress.gov/109/plaws/publ58/PLAW-109publ58.pdf
14. Le A, Rodrigo N, Domingo N, Senaratne S (2023) Policy mapping for net-zero-carbon buildings: insights from leading countries. Buildings 13(2766):1–23. https://doi.org/10.3390/buildings13112766
15. The US Government (2021) The long-term strategy of the United States: pathways to net-zero (issue Nov). https://www.whitehouse.gov/wp-content/uploads/2021/10/us-long-term-strategy.pdf
16. Fam A, Fam S (2024) Review of the US 2050 long term strategy to reach net zero carbon emissions. Energy Rep 12(June):845–860. https://doi.org/10.1016/j.egyr.2024.06.031
17. Center for Climate and Energy Solutions (2024) U.S. state climate action plans. https://www.c2es.org/document/climate-action-plans/
18. Elkind EN, Lamm T, Segal K (2020) Capturing opportunity: law and policy solutions to accelerate engineered carbon removal in California. http://www.climatepolicysolutions.org/wp-content/uploads/2021/03/Capturing-Opportunity-December-2020-1.pdf
19. Regional Greenhouse Gas Initiative (2024) The regional greenhouse gas initiative. An Initiative of Eastern States of US. https://www.rggi.org/
20. Kutty N (2023) How Japan is accelerating efforts towards a carbon-neutral society. World economic forum annual meeting. https://www.weforum.org/agenda/2023/01/davos23-japan-accelerate-efforts-carbon-neutral-society/
21. The Government of Japan (2024) Together for action: Japan's initiatives for achieving the common goal of net zero by 2050
22. Kuwabara T, Mohr D, Sauer B, Yamada Y (2021) How Japan could reach carbon neutrality by 2050. https://www.mckinsey.com/capabilities/sustainability/our-insights/how-japan-could-reach-carbon-neutrality-by-2050
23. Ministry of Economy Trade and Industry (METI) (2020) Environment innovation strategy. https://unit.aist.go.jp/gzr/zero_emission_bay/en/images/kankyousenryaku2020_english.pdf
24. Ministry of Economy Trade and Industry (METI) (2021) Overview of Japan's green growth strategy through achieving carbon neutrality in 2050. https://www.mofa.go.jp/files/100153688.pdf
25. Blaise Hope (2022) Japan's route to its 2050 net and 'beyond zer' targets. Sustainability Magazine 1. https://sustainabilitymag.com/net-zero/japans-route-to-its-2050-net-and-beyond-zero-targets-emissions-cop26

26. Ministry of Economy Trade and Industry (METI) (2023) Green innovation fund. https://www.meti.go.jp/english/policy/energy_environment/global_warming/gifund/index.html
27. Obayashi Y, Golubkova K (2023) Explainer: Japan carbon pricing scheme being launched in April. Carbon Markets. https://www.reuters.com/markets/carbon/japans-carbon-pricing-scheme-being-launched-april-2023-03-30/
28. Appunn K, Eriksen F, Wettengel J (2024) Germany's greenhouse gas emissions and energy transition targets. J Energy Transit
29. Wettengel J (2024) Germany to adopt 2060 target for net-negative greenhouse gas emissions. J Energy Transit. https://www.cleanenergywire.org/news/germany-adopt-2060-target-net-negative-greenhouse-gas-emissions
30. The Federal Government (2024) Climate action programme 2030. Climate Action
31. Federal Climate Change Act 1 (2021) Testimony of and nuclear safety (BMUV). Federal Ministry for the Environment, Nature Conservation
32. International Energy Agency (IEA) (2023) Federal Climate Change Act 2021. Policies. https://www.iea.org/policies/13518-federal-climate-change-act-2021
33. Meza E (2024) German special climate fund remains untapped to large extent. Clean Energy Wire
34. International Energy Agency (IEA) (2019) Policies, p 1. https://www.iea.org/policies/863-building-energy-efficiency-act
35. Green Building Construction Support Act 1 (2012) Testimony of statues of the Republic of Korea. https://elaw.klri.re.kr/eng_mobile/ganadaDetail.do?hseq=46773&type=abc&key=GREENBUILDINGSCONSTRUCTIONSUPPORTACT&param=G
36. Joo J, Paavola J, Van Alstine J (2023) Contested net-zero target setting in a transitioning country: the case of South Korea. Futures 147(103114):1–12. https://doi.org/10.1016/j.futures.2023.103114
37. International Energy Agency (IEA) (2020) Energy policy review Korea 2020
38. Government of Canada (2024) Net zero emissions by 2050. Climate change: our plan. https://www.repsol.com/en/sustainability/climate-change/net-zero-emissions-2050/index.cshtml
39. Halifax NS (2024) Government of Canadas new Canada green buildings strategy to help Canadians save money on their energy bills. Innovation, Science and Economic Development Canada. https://www.canada.ca/en/innovation-science-economic-development/news/2024/07/government-of-canadas-new-canada-green-buildings-strategy-to-help-canadians-save-money-on-their-energy-bills.html
40. Canadian Net-zero Emissions Accountability Act, Justice Law 22 (2021). https://laws-lois.justice.gc.ca/eng/acts/c-19.3/fulltext.html
41. The Senate of Canada (2023) A short guide to the Canadian net-zero emissions accountability act (CNZEAA). Senator Rosa Galvez. https://rosagalvez.ca/en/initiatives/climate-accountability/short-guide-to-the-cnzeaa/
42. Haq A, Sharma R, Bakshi BR, Kodamana H, Ramteke M (2023) Forecasting sustainable power generation profiles to achieve net zero emissions using multi-objective techno-ecological framework: a study in the context of India. Comput Chem Eng 179(Sept):108439. https://doi.org/10.1016/j.compchemeng.2023.108439
43. Feng W, Berkeley L (2016) Impact analysis of developing net zero energy buildings in China, pp 1–14
44. Singhania M, Saini N (2023) Institutional framework of ESG disclosures: comparative analysis of developed and developing countries. J Sustain Finance Investment 13(1):516–559. https://doi.org/10.1080/20430795.2021.1964810
45. Zhang SC, Yang XY, Xu W, Fu YJ (2021) Contribution of nearly-zero energy buildings standards enforcement to achieve carbon neutral in urban area by 2060. Adv Clim Chang Res 12(5):734–743. https://doi.org/10.1016/j.accre.2021.07.004
46. Li J, Gou Z (2024) Addressing the development gap in net-zero energy buildings: a comparative study of China, India, and the United States. Energy Sustain Dev 79(Feb):101418. https://doi.org/10.1016/j.esd.2024.101418

47. Chandel R, Chandel SS, Prasad D, Dwivedi RP (2024) Sustainable passive solar and photovoltaic integrated technology interventions for climate responsive net zero energy buildings in western Himalayan mountainous terrain of India. Next Sustain 3(Nov 2023):100039. https://doi.org/10.1016/j.nxsust.2024.100039
48. Bureau of Energy Efficiency (BEE) (2018) Energy conservation building code for residential buildings. Ministry of Power, Government of India, vol 1. https://www.beeindia.gov.in/sites/default/files/ECBC_BOOK_Web.pdf
49. Panicker K, Anand P, George A (2023) Assessment of building energy performance integrated with solar PV: towards a net zero energy residential campus in India. Energy Build 281:112736. https://doi.org/10.1016/j.enbuild.2022.112736
50. Das A, Saini V, Parikh K, Parikh J, Ghosh P, Tot M (2023) Pathways to net zero emissions for the Indian power sector. Energy Strategy Rev 45(Aug 2022):101042. https://doi.org/10.1016/j.esr.2022.101042
51. The Asia Society (2023) Getting Indonesia to net zero, pp 1–14
52. Svendsen A (2022) Roadmap for an energy efficient, low-carbon buildings and construction sector in Indonesia. Global alliance for buildings and construction, pp 1–95
53. Pattinggi DMM, Goerke M, Gadd A (2021) Introducing Indonesia's new carbon law. https://www.herbertsmithfreehills.com/notes/indonesia/2021-12/introducing-indonesias-new-carbon-law
54. Wahyuni E, Ardiansyah H (2022) Indonesia's national strategy and commitment towards transition to renewable energy. In: Indonesia post-pandemic outlook: strategy towards net-zero emissions by 2060 from the renewables and carbon-neutral energy perspectives, pp 9–22. https://doi.org/10.55981/brin.562.c2
55. Saudi Green Initiative (2024) Proactive green initiatives for positive global impact striving to achieve net zero in Saudi Arabia by 2060. Saudi & Middle East Green Initiatives
56. Kamboj P, Hejazi M, Alhadhrami K, Qiu Y, Kyle P, Iyer G (2023) Saudi Arabia net zero GHG emissions by 2060 transformation of the electricity sector (issue Dec). https://doi.org/10.30573/KS--2023-DP31
57. Saudi Energy Efficiency Center (2023) Saudi energy efficiency center: building sector. Saudi Energy Efficiency Center
58. Matar W, Mansouri NY, Umeozor E (2024) Energy policy pathways to inform climate policy in Saudi Arabia. J Clean Prod 445(Feb 2024):141190. https://doi.org/10.1016/j.jclepro.2024.141190
59. Ministry of Energy Saudi Arabia (2024) National renewable energy program. National Renewable Energy Program
60. Circular Carbon Economy Initiative (2024) Circular carbon economy national program. Asia Indigenous Peoples Pact
61. Crippa M, Guizzardi D, Banja M, Solazzo E, Schaaf E, Pagani F, Monforti-Ferrario F, Olivier JJG, Quadrelli R, Risquez Martin M, Moharamli Taghavi P, Grassi G, Rossi S, Oom D, Branco A, San Miguel J, Vignati E, Muntean M, Schaaf E, Vignati E (2022) Fossil CO_2 emissions of all world countries—2020 report. JRC Publications Repository. https://doi.org/10.2760/07904
62. Engel H, Helmcke S, Ruth H, Hieronimus S (2021) Net-zero Germany: chances and challenges on the path to climate neutrality by 2045. McKinsey & Company
63. Li B, Haneklaus N (2022) The potential of India's net-zero carbon emissions: analyzing the effect of clean energy, coal, urbanization, and trade openness. Energy Rep 8:724–733. https://doi.org/10.1016/j.egyr.2022.01.241
64. United Nations Environmental Programme (UNEP) (2023) Temperatures hit new highs, yet world fails to cut emissions again. In: Emissions gap report 2023. https://doi.org/10.59117/20.500.11822/43922
65. Park BR, Chung MH (2023) Analysis of the additional energy-saving potential of residential buildings after mandatory zero-energy buildings to achieve carbon neutrality in South Korea. Build Environ 228(2023):1–10. https://doi.org/10.1016/j.buildenv.2022.109908

Stakeholders and Their Roles in the Net Zero Carbon Construction Industry

Coralie van Reenen, Jeremy Gibberd, Tobias van Reenen, A. K. J. Surridge, Lisa Reynolds, Laetitia van der Merwe, Jan Hugo, Megan Euston-Brown, and Bernd Oellermann

Abstract This chapter investigates the role of stakeholder collaboration and engagement in achieving net zero carbon (NZC) buildings, which are essential to mitigate climate change. Barriers to the effective implementation of NZC buildings are identified, and the contribution of stakeholders in overcoming these is discussed, specifically with reference to engaging local communities and environmental groups, preparing the workforce, and education and training. The chapter draws on the insights of stakeholders collected through a survey and presents success stories that showcase effective collaboration in achieving sustainable construction goals. It applies stakeholder theory to the construction industry, identifying four categories of stakeholders in the value chain to achieve NZC buildings, namely policy makers, industry actors, end users, and civil society. The roles of these stakeholders and the importance of coordination and collaboration at an industry-wide level are discussed. The chapter identifies key factors that are necessary for successful industry transformation and recommends that strong leadership and clear objectives are necessary to facilitate a coordinated effort that produces change.

C. van Reenen (✉) · J. Gibberd · T. van Reenen · B. Oellermann
Council for Scientific and Industrial Research, Pretoria, South Africa
e-mail: cvreenen@csir.co.za

A. K. J. Surridge
South African National Energy Development Institute, Sandton, South Africa

L. Reynolds
Green Building Council of South Africa, Cape Town, South Africa

L. van der Merwe · J. Hugo
University of Pretoria, Pretoria, South Africa

M. Euston-Brown
Sustainable Energy Africa, Cape Town, South Africa

N. Domingo et al. (eds.), *Global Net Zero Carbon Practices in Construction*, Green Energy and Technology, https://doi.org/10.1007/978-981-96-3853-6_4

1 Introduction

The construction industry is one of the major contributors to greenhouse gas emissions and climate change, accounting for 38% of global energy-related carbon dioxide emissions. To achieve the global goal of limiting the temperature rise to 1.5 °C above pre-industrial levels, the construction industry must reduce its emissions by 80% by 2050, relative to 2010 levels. Stakeholder collaboration in the construction industry is critical to achieving this goal. This chapter seeks to understand the different roles that stakeholders in the built environment can play in achieving net zero carbon (NZC) goals.

Radical transformation of the built environment industry is required to meet NZC targets so that buildings contribute zero or negative carbon emissions, including carbon off-sets, over their life cycle.[1] Such a radical transformation in any industry comes with challenges, such as economic barriers, market limitations, lack of knowledge, technical challenges, and organisational and legal constraints [13]. Stakeholder collaboration has been identified as a key strategy to overcome some of these challenges [19].

The purpose of stakeholder collaboration in achieving NZC buildings is to create a shared vision and commitment among all parties involved in the construction industry and to align their interests, values, expectations, and actions, while promoting innovation, knowledge exchange, and learning, and reducing conflicts, risks, and costs. Inclusive and participatory engagement of stakeholders can lead to NZC construction becoming more financially feasible and socially acceptable.

The different role-players involved in lifecycle stages and the extensive value chain of construction and buildings mean that there are many stakeholders to consider. To achieve NZC, these stakeholders must be identified, their activities mapped, and their potential contribution to enable the achievement of NZC understood. This research aims to provide a framework that can be used for this analysis and to enable stakeholder coordination.

The chapter considers the literature on the role of stakeholders in achieving NZC buildings, establishing four categories of stakeholders, and explores their roles in engaging local communities, preparing the workforce, and training and education. The findings are analysed and discussed to establish not only the potential role of each stakeholder, but also their relationships and power dynamics. This provides a clear understanding of the stakeholder network so that the NZC building goals can be established and implemented with their full support.

---

[1] For the purpose of this discussion, net positive carbon construction is considered to be included in NZC.

2 Landscape Review

2.1 Stakeholder Theory

Framing stakeholder theory in the context of business management could help align the NZC goal with commercial interests as opposed to a pure regulatory imposition. In business management, stakeholder theory and stakeholder engagement refer to business interactions between stakeholders, such as clients, employees, suppliers, and other parties in the value chain providing a service or a product [11]. In the context of the NZC goal, 'business' or 'product' is a target industry norm, constituted by the successful implementation of desirable NZC solutions with informed and satisfied partners.

The principle behind stakeholder theory is that the success of an endeavour relies on effective management of its interactions with groups that can influence the fulfilment of its purpose [8]. Stakeholders are defined as any party that affects or is affected by the achievement of and the purpose [7]. In the context of achieving NZC building targets, the range of parties that fit this definition is very broad, considering the extensive upstream and downstream value chains in the life cycle of a building.[2]

2.2 Construction Industry Stakeholders

The life cycle of a building can be divided into five stages namely preconstruction, construction, operation and maintenance, demolition, and reuse [6]. At each stage, several stakeholders can be identified. Falana, et al. [6] refer to NZC building stakeholders as 'individuals or groups of individuals who are interested in the development of net zero carbon building and are willing to utilise their skills, experience, knowledge, and resources in achieving a net zero carbon building whole life cycle'—this definition limits itself to those parties who support the concept of NZC. However, by Freeman's definition, those who do not support NZC construction, for whatever reason, are still stakeholders. This positive and negative influence of stakeholders emphasises the need for stakeholders to engage with each other throughout the value chain to align their values and objectives.

Insufficient stakeholder engagement is cited as a challenge to the realisation of NZC buildings and suggests that delivery of such buildings has social, cultural, and political implications, making this a socio-technical matter [2]. The scope of affected parties indicates that a broader range of stakeholders should be involved, not only in individual building projects, but in the industry-wide conversation to transform the industry.

A collaborative and participatory approach can facilitate the exchange and integration of information and knowledge among stakeholders, as well as the co-creation and

[2] https://inclusiveinfra.gihub.org/action-areas/project-planning-development-and-delivery/.

co-delivery solutions. Furthermore, a collaborative and participatory approach can foster trust, commitment, and shared ownership among stakeholders and empower them to collectively act for change.

Several methods and tools can be used to support a collaborative and participatory approach, such as stakeholder mapping and analysis, stakeholder engagement workshops, focus groups, surveys, interviews, design charrettes, living labs, pilot projects, feedback mechanisms, and monitoring and evaluation systems [6, 20]. Although there is a significant body of literature on stakeholders and their roles in achieving net zero carbon buildings [6], these tend to focus on specific examples or projects, rather than the entire industry. Although individual projects play a role in gradually turning the industry direction, a broader, industry-wide collaborative effort is needed to significantly and quickly shift the industry norm.

2.3 *Barriers and Enablers to Implementation of NZC Buildings*

2.3.1 Comprehensive Communication

The literature highlights some of the common barriers and challenges that the uptake of NZC buildings poses, one of which is a lack of awareness and knowledge among stakeholders, which hinders their understanding and adoption of net zero carbon construction [20]. For example, project owners and developers may lack the awareness of the market potential and financial benefits of NZC buildings, or the knowledge of how to incorporate them into their project briefs and tenders. Architects and engineers may lack access to design tools and optimisation methods for NZC buildings or familiarity with the relevant codes and standards. Contractors and suppliers may lack the skills and expertise to construct and equip NZC buildings, or the availability of suitable materials and technologies, or the drive for the lowest costs may dominate decision-making. Users and occupants may lack comprehension of the environmental and social impacts of their behaviour and choices, the skills to operate and manage net zero carbon buildings efficiently, or they may have a perception that the cost of NZC buildings is prohibitive.

This can be overcome through education and training to increase awareness and knowledge of the benefits and implications of NZC construction and equip them with the necessary skills and competencies to implement it [24].

2.3.2 Commercial Competition

Another obstacle is the perceived cost of NZC buildings, along with a lack of incentives and regulations. Many stakeholders do not have sufficient commercial, ethical, or regulatory incentives to motivate and guide their actions towards NZC construction. For example, government and policy makers may not have a clear vision and strategy for a NZC construction industry and may also fail to provide adequate incentives, such as subsidies, tax breaks, grants, loans, awards, certification, labelling, or disclosure. Financiers and investors may not have the appropriate criteria and indicators to evaluate the performance and risks of NZC buildings or may not have access to reliable and transparent data and information. Customers and users may not have adequate options or incentives to choose NZC buildings or may not be willing to pay a premium for them. Communities and NGOs may not have enough voice and influence to advocate for NCZ construction or may face resistance and opposition from competing stakeholders [16].

Incentives and regulations can provide the motivation and guidance for stakeholders to act towards net zero carbon construction. For example, government and policy makers can set and enforce clear and consistent targets and standards for NZC construction, such as codes, regulations, legislation, or policies. They can also provide financial and non-financial incentives, such as subsidies, tax breaks, grants, loans, awards, and certification or labelling. Financiers and investors can develop and apply criteria and indicators to assess and value the performance and risks of net zero carbon buildings, such as ratings, benchmarks, metrics, or indices. Customers and users can express and exercise their options and preferences for net zero carbon buildings, such as demand, choice, feedback, or recommendation.

2.3.3 Collaboration and Coordination

In addition to the above barriers, many stakeholders do not have effective collaboration mechanisms to facilitate and coordinate their interactions with respect to NZC construction. For example, project owners and developers may not engage and consult with other stakeholders at an early stage of the project to establish clear and mutual expectations and responsibilities. Architects and engineers may not collaborate and integrate their designs or ineffectively communicate and document them. Contractors and suppliers may not coordinate and align their delivery and installation processes or provide adequate feedback and quality control. Users and occupants may not participate and contribute to the operation and maintenance of NZC buildings or share their experiences and best practices.

Communities and NGOs can advocate and support net zero carbon construction, such as campaigns, petitions, protests, or partnerships.

Collaboration and communication can facilitate and coordinate the interactions and relationships of stakeholders towards net zero carbon construction and resolve conflicts among stakeholders. For example, project owners and developers can engage and consult with other stakeholders at an early stage of the project and establish clear and mutual expectations and responsibilities. They can also create and maintain a collaborative and supportive project culture and environment, such as vision, mission, values, or goals. Architects and engineers can collaborate and integrate their designs and solutions, and communicate and document them properly. They can also use and share the latest tools and methods to design and optimise net zero carbon buildings, such as software, models, simulations, or data. Contractors and suppliers can coordinate and align their delivery and installation processes and provide adequate feedback and quality control. They can also source and use the required materials and technologies for net zero carbon buildings, such as low-carbon, renewable, recycled, or local. Users and occupants can participate and contribute to the operation and maintenance of net zero carbon buildings and share and learn from their experiences and best practices. They can also monitor and measure the performance and impact of NZC buildings, such as energy, water, waste, or emissions.

2.4 *Precedents from Other Industries*

A global commitment to achieve net zero carbon buildings by 2050 can be likened to the global commitment to reduce CFCs in that both relate to changes needed globally to achieve important environmental changes. Although the reduction of CFCs in manufacturing affected a smaller sector of stakeholders, much can be learnt from the successful implementation of the initiative. Universal international agreement to ban CFCs was achieved through the Montreal Protocol; likewise, the goal of achieving net zero carbon buildings by 2050 was established as part of the broader goals of the Paris Agreement. Differences lie in the timelines set and the scope of change required. While it took only four years from the discovery of the hole in the ozone layer to the effective banning of CFCs [22], net zero building targets are set over a 20–30 year period, and the scope impacts both new and existing buildings of all occupancy types.

Even before it was legislated, the use of aerosol sprays was reduced based on the basis of campaigning by environmental groups. Within two years of the Montreal Protocol, CFCs were banned. This highlights the significant role the environmental groups and non-government organisations can play in shifting the industry.

The key strategies that led to the successful reduction in the use of CFCs include international cooperation, public awareness campaigning, industry innovation (providing alternatives), and regulatory measures. It is important to note that each of these strategies in isolation could not achieve the change required, and therefore, a collaborative and coordinated effort is required.

3 Summary and Discussion of Landscape Review

The landscape review shows that stakeholders have a vital role in achieving net zero carbon in the construction industry, but they also face various challenges in engaging among themselves towards this purpose, such as:

- Lack of clear and consistent definitions and standards for NZC construction, which can create confusion and inconsistency among stakeholders.
- Lack of sufficient and reliable data and information on the costs, benefits, and impacts of NZC construction, which can hinder decision-making and evaluation.
- Lack of adequate and accessible incentives and support for NZC construction, which may discourage investment and innovation.
- Lack of effective and inclusive communication and collaboration platforms and mechanisms, which can limit the coordination and integration of stakeholders.
- Lack of awareness and education about the importance and feasibility of NZC construction, which can reduce the demand and acceptance of stakeholders.

The review also suggests some opportunities and strategies to overcome these challenges and improve the engagement and alignment of stakeholders, such as:

- Developing and harmonising the definitions and standards for NZC construction, which may increase the clarity and consistency among stakeholders.
- Collecting and sharing data and information on the costs, benefits, and impacts of NZC construction, which can improve the decision-making and evaluation.
- Providing and promoting incentives and support for NZC construction, which may encourage investment and innovation.
- Creating and facilitating communication and collaboration platforms and mechanisms, which may enhance the coordination and integration of stakeholders.
- Raising awareness and education on the importance and feasibility of NZC construction, which can increase the demand and acceptance of stakeholders.

There is a lack of practical guidance on how to implement these opportunities and strategies effectively and efficiently and how to ensure collaboration among all stakeholders simultaneously at an industry level. Therefore, further research is needed to fill this gap and provide concrete and actionable recommendations for stakeholder engagement and alignment in net zero carbon construction.

4 Research Methodology

The methodology followed a collaborative framework of interdisciplinary research [25], through which the diverse ideas and experiences of contributors for different disciplines were incorporated. Contributors from various stakeholder sectors each provided their experiences and insights regarding their role in the global mission to achieve net zero carbon buildings. Sectors represented include government agencies, research institutes, non-government organisations, industry professionals, and academia. All have been involved in various initiatives to promote and support the objective of achieving net zero carbon buildings by 2050.

The objective of the research was to explore stakeholder roles and establish a method to unite and consolidate stakeholder efforts to accelerate the transition of the construction industry to net zero carbon.

Within the collaborative framework, an exploratory research methodology was followed for the study. Exploratory research is suitable when not enough is known about an area and when key issues have not been clearly defined [10, 15]. It provides a basis for the generation of ideas and hypotheses for further in-depth research. The methodology is suitable for understanding human behaviour, experience, attitudes, intentions, and motivations and has been used to understand the different roles of stakeholders in relation to achieving change [3, 14].

The methodology has been selected for this study as there has been limited research on the coordination of stakeholders' roles in supporting net zero policies in countries to ensure that sectors, such as the building and construction sectors, help achieve the National Determined Contributions (NDCs) [17]. For efficient implementation, key stakeholders will need to be identified, their roles understood, and their activities coordinated and integrated. This study aims to contribute to knowledge within this area.

It investigates the role of key stakeholders in achieving net zero targets in the building and construction sector. The contributors are all from the South African built environment sector; despite this context, the discussions are not limited to South Africa and the findings are universally applicable to developed and developing countries.

Research will include both primary and secondary research in the form of a literature review. The primary research consisted of an online survey of contributors. The survey asks the following questions:

- Which one of the following categories of stakeholders best describes your work/role/influence in the context of the built environment industry?
- In this category, what role do you think you could / should / do play in engaging with local communities and environmental groups to support achieving net zero carbon buildings? Please elaborate with examples, if you can.
- In this category, what role do you think you could/should/do play in preparing the workforce (those already in the industry and/or those preparing to enter the industry)?
- In this category, what role do you think you could / should / do play in education and training?
- Can you provide any perspectives and/or success stories that could help the industry transition towards net zero carbon buildings effectively?

Data from the survey are collected, presented, and analysed, synthesising the inputs into a discussion of stakeholder roles. To understand the results, the study draws on secondary research to discuss the findings. In particular, secondary research will be used to define stakeholder roles more clearly and understand how they may contribute to the achievement of net zero energy targets. The study does not aim to achieve conclusive results and recommendations. Instead, it aims to provide a framework and basis from which further research and work in this field can be carried out.

5 Findings and Discussion

The findings, based on input from various contributors, are discussed in this section.

5.1 The Role of Stakeholder Collaboration and Engagement

It is clear that involving stakeholders is crucial for shifting the built environment towards net zero carbon. Engaging a broad range of people and groups brings diverse knowledge, views, and resources to the goal. A primary facilitator of this engagement needs to be identified to coordinate stakeholder roles. Even with net zero carbon policies in place, their execution remains largely uncoordinated, with isolated pockets of success.

Pan and Pan [20] identify four components of the industry landscape that influence the uptake of NZC buildings, namely supply chain, policy and regulation, technology and innovation, and user behaviour, placing policy and regulation at the centre, as shown in Fig. 1. By doing so, it is implied that policy and regulation are the main drivers for shifting the industry, which is a top-down approach. While policy and enforcement are essential components, this strategy bears the risk of leaving others to feel like the victims of a policy, rather than partners in a journey.

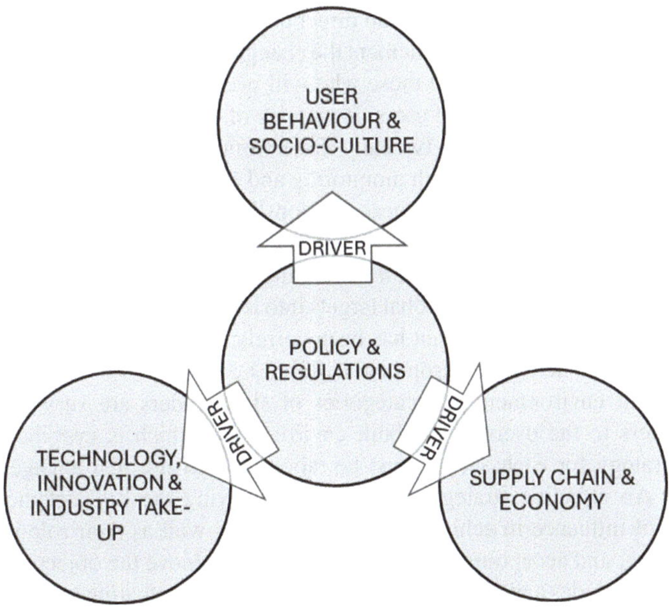

Fig. 1 Socio-technical systems framework of delivering zero carbon buildings

The main driver of policy and regulation is the global commitment to reduce greenhouse gas emissions, as has been established through various international frameworks and agreements, such as the Paris Agreement, the 2030 Agenda for Sustainable Development, and the New Urban Agenda.

These commitments materialise through national policies and regulations that provide the legal and institutional basis for implementing net zero carbon strategies. These in turn filter down to local governments and the private sector for implementation. For example, the UK has adopted the Climate Change Act 2008, which sets a legally binding target of reducing its greenhouse gas emissions by 80% by 2050 compared to 1990 levels [5]. In turn, the Code for Sustainable Homes in the UK was established as a voluntary rating system that assessed the environmental performance of new homes and awarded credits for achieving higher standards of performance [4].

In South Africa, the National Development Plan (NDP) includes a NZC goal for buildings by 2030. Despite there being no implementation strategy at a national level, some municipalities have established mechanisms for achieving the target and the private sector also drives NZC buildings through their ESG reporting requirements. Whether the driver is national policy, municipal strategies or private sector incentives, there is need for stakeholder engagement and an established process for collaboration.

The implementation of net zero carbon buildings is not a one-off event, but a dynamic and iterative process that requires constant feedback and learning. There are several tiers of stakeholders: at the upper end of the hierarchy, are those stakeholders who are demanding change and have authority over the system, such as government departments; there are those who direct the change, informed by research and standards; there are those who implement the change, such as architects and construction contractors; lastly, there are those who will need to accept the change and use or operate the product—the end users. If any one of the tiers is missing, the system collapses. While the imperative typically flows from the top, the flow of communication needs to go both ways, with monitoring and evaluation and feedback loops to ensure continuous improvement. These relationships are illustrated in Fig. 2.

The Graz Declaration for Climate Protection in the Built Environment highlights some of the relevant stakeholders in transitioning the industry. The overall objective of the declaration is to translate global targets into actions for specific sectors, leading to five key points [21]. Each point has been correlated with relevant role-players in the value chain of the built environment in Table 1.

In the built environment, the categories of stakeholders are very broad, from policy makers to the users of the built environment (which is everybody), and a suitable strategy for each sector must be tailored to ensure that engagements are impactful. An effective strategy starts with identifying the key stakeholders and their level of influence in achieving the objectives as well as their role in guiding, implementing, and accepting the actions required to achieve the objectives. A good strategy also includes a methodology for monitoring and evaluating progress towards the objectives.

The NZC construction industry involves a complex network of stakeholders, who, for the sake of simplicity, can be classified into four categories: policy makers,

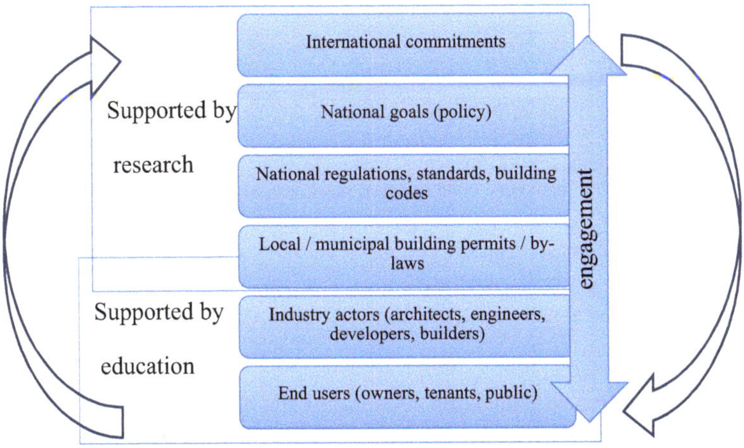

Fig. 2 System of stakeholders

Table 1 Key points of Graz Declaration relative to relevant stakeholders

Graz Declaration point Passer et al. [21]	Relevant role-players
Setting national design targets for the carbon footprint of buildings over their life cycle through standardisation and legislation	Legislators, policy makers, government departments, international organisations, standard setters
Recognising that operational and embodied carbon are to be accounted for	Material suppliers, designers, building operators, building maintainers
Recognising the role of the research community in establishing science-based targets to be included at policy level, as well as advice and support industry and society in implementation	Academia, research councils
Developing appropriate new skills and capabilities through higher education and training	Academia, tertiary institutions, professional bodies, non-government organisations, and lobby groups
Recognising the need for continuous upskilling and training within the built environment professions and existing workforce	Professional bodies, training institutes, non-government organisations, and lobby groups

industry actors, end users, and civil society. Each category has different subgroups with specific roles, as shown in Table 2.

Each stakeholder has a different role to play and level of engagement at different stages of the building life cycle and value chain (shown in Fig. 3). Naturally, policy makers have a strong role in setting policies and regulations that govern the value chain and influence the whole system. Industry actors have an influence in the planning and design, material choices, construction, and financing. End users influence the design to an extent and the operational phase. Civil society has a cross-cutting

Table 2 List of stakeholder categories

Category	Subgroups
Policy makers	Government agencies, regulators, standard setters, international organisations
Industry actors	Developers, investors, financiers, architects, engineers, quantity surveyors, construction managers, contractors, consultants, suppliers, manufacturers, facility managers
End users	Owners, tenants, occupants, customers, clients
Civil society	Media, academia, research institutions, non-governmental organisations, community groups, general public

role in creating awareness, educating and guiding the industry actors, end users, and policy makers.

Once the groups of stakeholders and their roles in the value chain are understood, it is necessary to examine their roles in relation to one another in the activities of engaging with local communities and environmental interest groups, preparing the workforce, and education and training, as discussed in the following sections.

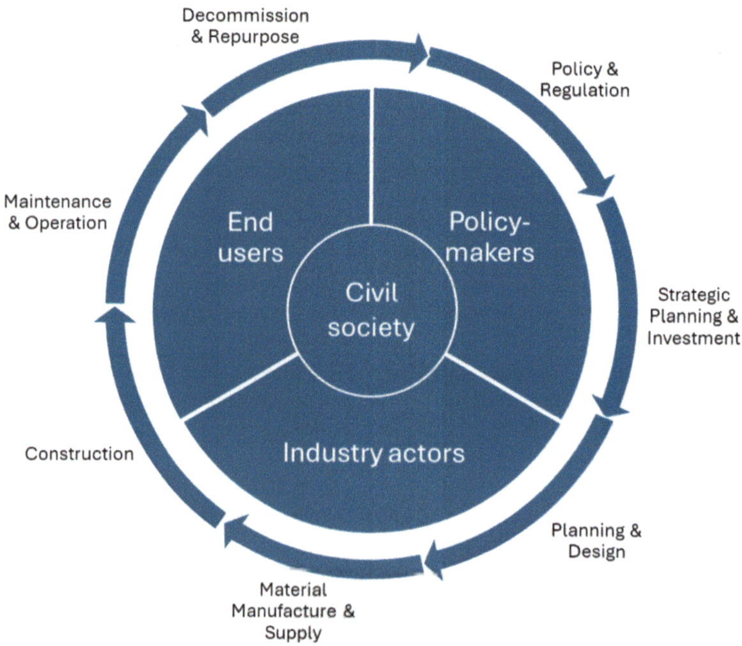

Fig. 3 Life cycle stages of a building

5.2 Engaging with Local Communities and Environmental Groups

Local communities and environmental groups fall within the stakeholder categories of civil society and end users and include residents, businesses, faith groups, school communities, and neighbourhood syndicates, green building councils, and other non-government organisations (NGOs).

These groups can influence the construction industry, advocating for NZC policies and supporting the implementation. This influence, as well as their role as end users, makes local communities and environmental groups essential stakeholders; suitable engagement with them is necessary to support industry-side uptake of NZC building practices. The role of each category of stakeholders with specific reference to local communities and environmental groups is discussed below.

5.2.1 Policy Makers' Role in Engaging with Local Communities and Environmental Groups

Policy makers, such as national departments, local authorities, and regulators, need to engage with local communities and environmental groups as resources that provide input on policies, ensuring that they are relevant, effective, and responsive to the needs and challenges of different contexts. Policy makers can benefit from the expertise, experience, and feedback of these groups, who are often at the forefront of innovation and advocacy for NZC buildings. Local communities and environmental groups can also provide valuable insights and feedback on the social, cultural, and ecological aspects of NZC.

The research survey found that policy makers, such as regulatory bodies[3] and local government agencies,[4] believe that they have a role in facilitating dialogue and knowledge exchange through workshops. It is not possible for policy makers to meet with all local communities and environmental groups. However, by engaging with key representatives from these groups, these can then become agents disseminating knowledge through workshops with other groups.

For example, the World Green Building Council (WGBC) is a global network of green building councils that promotes the adoption of NZC standards and practices in the building sector. In South Africa, the Green Building Council (GBCSA) has formed good relationships with policy makers, such as the South African Local Government Association (SALGA), individual municipalities, and the National Home Builders Regulatory Council (NHBRC) and regularly provides input on policy and implementation upon their request. The NGO, Sustainable Energy Africa (SEA), promotes equitable, low carbon, clean energy development in South Africa and Africa

[3] NHBRC.

[4] SALGA.

through research, capacity building, policy engagement, and information dissemination and works extensively with municipalities to implement NZC strategies, where the municipalities lack capacity to do so themselves.

5.2.2 Industry Actors' Role in Engaging with Local Communities and Environmental Groups

Unlike the policy makers, industry actors are typically involved in transitioning the industry in a very practical way by designing, building, and operating NZC buildings on a project-by-project basis. At a project level, these industry actors often need to spend resources convincing end users and recipient communities of the benefits of a NZC building.

Through the research survey, it was reported[5] that in private sector projects, the activity of negotiating, defending, and motivating for NZC buildings is as an additional service with an additional cost and is often trimmed from the project. To ensure that this essential step is not neglected, environmental groups or local community leaders can be used to fulfil this role on behalf of the project.

It is also important for industry actors to learn from the community, making it a bidirectional engagement. To achieve NZC, the designers and project leaders need to learn what local community systems are available and sustainable and make use of the most contextually appropriate solutions. It is important for a structured methodology to be established, perhaps through industry professional bodies and environmental groups, to guide architects and project leaders in executing successful community engagements.

5.2.3 End Users' Role in Engaging with Local Communities and Environmental Groups

End users are those who occupy or own buildings. They need to engage with local communities and environmental groups to learn about NZC buildings. For owners and project clients, this is important as they approve the design and costs of buildings.

End users are not always aware of NZC buildings and there is no formal system to educate them. While industry actors, community groups, or environmental groups can inform them at a project level, there is a gap for creating awareness and educating the wider industry and public. Local community groups and environmental groups can play this role, but they need support from an invested stakeholder. A coordinated effort and leadership are required.

[5] MSSA, Marc Sherratt.

5.2.4 Civil Society's Role in Engaging with Local Communities and Environmental Groups

While community groups and environmental organisations are included in the civil society stakeholder group, it is also essential to take into account the contribution of other significant entities within civil society, like academic and research institutions.

The survey showed that some researchers report feeling disconnection from local communities and environmental organisations, contributing only indirectly through their scholarly and popular articles, policy recommendations, and thought leadership. Academics are tasked with generating and disseminating knowledge and play a significant role in supporting and verifying the role of environmental groups. In South Africa, for example, the Council for Scientific and Industrial Research (CSIR), as well as many of the universities, have a healthy knowledge-sharing relationship with the GBCSA, which has a leading influence on industry actors and policy makers.

This relationship and continuous bidirectional knowledge sharing needs to be nurtured. The academic and research community needs to ensure that knowledge disseminated to and by local communities and environmental groups is accurate, and 'greenwashing' is avoided. Communication and engagements can occur through workshops, round table meetings, conferences or living laboratories. Living laboratories provide an intersection of research and local communities, utilising a participatory approach involving multiple stakeholders in co-creating solutions for complex sustainability challenges [9]. A living laboratory can demonstrate best practices and innovative technologies for NZC buildings, and serve as a testbed for evaluating their performance and user satisfaction [12].

The main finding regarding the role of stakeholder groups in engaging with local communities and environmental groups is that these stakeholders are potentially active at every level of the industry, playing an advocacy and implementation support role to policy makers, civil society, industry actors, and end users.

5.3 Preparing the Workforce

The workforce refers to those who are already practising in industry and may never have been formally educated in the principles and practices of NZC buildings.

The workforce consists of crucial stakeholders, as it consists of the individuals who design, construct, operate, and maintain buildings. As the construction industry moves towards low-carbon practices, there is a growing need for the workforce to develop and maintain suitable capabilities [23].

5.3.1 Policy Makers' Role in Preparing the Workforce

Policies, regulations, and standards create a requirement for the workforce to transition to a NZC industry. Policy makers view their role in preparing the workforce for the transition as one of advocacy and capacity-building through workshops about the policies and regulations. As discussed in the previous section, policy makers can make use of community and environmental groups to assist with this task; however, it is noted that there is also tremendous value in national departments hosting industry workshops to engage directly with the workforce.

A reciprocal relationship is necessary between employees and government bodies aiming to achieve national goals; the workforce should be allowed to contribute insights, while also being ready for changes to the norm. Including the workforce in policy making is likely to smooth the enforcement process and minimise opposition.

5.3.2 Industry Actors' Role in Preparing the Workforce

Industry actors, such as contractors, developers, consultants, and suppliers, have a key role in engaging with the workforce, of which they are a part, to prepare them to move the industry towards NZC buildings targets.

Industry participants can equip their workforce for NZC buildings by offering mentorship programmes, including online training, workshops, site visits, case studies, and collaborative learning groups. The survey found that some industry actors are self-funding research and development activities to integrate net zero principles in their field and also mentor their employees.

Professional bodies have a significant role in providing continuous professional development courses, in collaboration with researchers, academia, and environmental groups. Communities of practice and knowledge-sharing platforms also provide opportunities for industry actors to build capabilities and 'normalise' the adoption of net zero building practices. This can help to create demand and awareness for NZC buildings and demonstrate their leadership and commitment to sustainability.

Upskilling and preparing for a changing industry helps actors gain a competitive edge, boost efficiency and profits, and uphold their social and ethical commitments.

5.3.3 End Users' Role in Preparing the Workforce

While the role of end users in preparing the workforce is not immediately apparent, they do have the potential to create a demand for NZC buildings. Such a client demand will stimulate the workforce to prepare themselves through CPD training.

5.3.4 Civil Society's Role in Preparing the Workforce

This group of stakeholders contributes significantly to upskilling the existing workforce. For example, SEA and the CSIR have developed and presented training to municipal building control officers regarding the approval of NZC building plans for construction. The GBCSA is very actively involved in training professionals on NZC building practices through workshops, courses, and online resources.

Architecture 2030, an international non-profit organisation, works with architects, engineers, designers, educators, and students to transform the built environment and address the climate crisis, providing tools, data, and guidance to help the building industry meet these targets and adopt low-carbon design practices [1].

The above discussion highlights that the civil society stakeholder group, including environmental groups, researchers, and academics, has a significant role in preparing the workforce.

5.4 *Education and Training in the Construction Industry*

Knowledge and skills development have been established as a key component of achieving NZC buildings [18]. While upskilling the workforce has already been addressed, this section looks at the role of various stakeholders in educating and training the future workforce at undergraduate level, including future policy makers, building users, property developers, building designers, construction managers, and quantity surveyors.

5.4.1 Policy Makers' Role in Education and Training

Policy makers have minimal engagement with students and emerging industry professionals, instead concentrating on current workforce interactions. However, it is essential to maintain open lines of communication with academic institutions and educators. Building robust connections can facilitate the exchange of knowledge and dismantle the apparent disconnect between policy makers and the student body.

5.4.2 Industry Actors' Role in Education and Training

Historically, the collaboration between industry professionals and academic institutions is limited. Yet, fostering robust partnerships between the two can yield great advantages. Precedents of success include cases where industry experts deliver guest lectures at universities, imparting real-world insights to students; additionally, companies provide internships that offer practical experience for students and supplementary support for the company's projects. Such alliances are advantageous for both parties.

5.4.3 End Users' Role in Education and Training

While end users do not typically have a role in education and training and are not necessarily recipients of NZC education as students, it is important to ensure that this group is suitably exposed to the benefits and implication of NZC buildings.

It is clear to see that academia has the biggest part in education and training of the future workforce, supported by research and environmental groups.

5.4.4 Civil Society's Role in Education and Training

Civil society, which among others includes academia, is key in education and training.

Educators must prepare the future workforce by including NZC building concepts in the curriculum. For example, at the University of Pretoria, module content in the Construction Economics Department covers green building principles, carbon tax, and carbon credits, as well as building performance optimisation using Building Information Modelling (BIM).

Education need not only be through traditional lectures. Deep learning can also be used as a teaching technique, focusing on interdisciplinary approaches [26]. Civil society can also participate in education and training through living laboratories, which include collaboration with all stakeholders. For example, the HSB Living Lab Building at Chalmers University was initiated as a collaboration between the university and twelve industry partners. The project also includes end users as part of the experiment, allowing for initiatives to test proof-of-concept products as well as research into the impact of behaviours and changes thereof.

5.5 *Success Stories and Lessons Learned*

Achieving net zero carbon buildings is notably challenged by insufficient coordination across the construction industry, from policy makers to consumers. This section shows success stories demonstrating that synchronised stakeholder involvement can promote the acceptance and application of NZC building ideas, methods, and technologies. It can also surmount obstacles hindering the shift to a low-carbon built environment. Furthermore, these instances underline the advantages of such collaboration, including innovation, knowledge exchange, cost savings, and social inclusion.

One success story that could be considered an example of collaboration and coordination across various stakeholders in the sector is the establishment of the National Eco-Industrial Park (EIP) Roundtable in South Africa. This forum, established in 2020, is convened every quarterly as an open platform, meaning anyone can participate. The first aspect of collaboration that needs to be in place is a Terms of Reference or Rules of Engagement that sets out the protocols for meetings and the objectives of the roundtable forum.

The second key aspect for a successful collaboration platform that can be taken from the EIP Roundtable is that discussion topics need to be relevant and up-to-date and sometimes be future-oriented and disruptive or provocative.

The third key aspect that ensures effective and efficient collaboration is strong facilitation and a convening power. In the case of the EIP Roundtable, this was achieved through institutional leadership of the Department of Trade, Industry, and Competition (the DTIC) and as a facilitator and convenor from the CSIR. This role requires an individual who is passionate and knowledgeable about the sector.

Lastly, collaboration and coordination can be enhanced through growing the base of participants continuously. This is achieved through extending the invitation to networks of networks to increase the reach.

The EIP Roundtable was initiated through a national department (policy maker), it is facilitated by members of a research organisation (civil society), is attended by engineers, architects and facility managers (industry actors), and end users, providing and example of interaction across the full spectrum of stakeholders. Replication of this method of engagement, which occurs through workshops for sharing ideas, could go a long way to unite the construction and buildings industry in transitioning to net zero carbon.

Another example of the results of good stakeholder coordination is found in the case of four major South African metropolitan municipalities—Cape Town, eThekwini, Johannesburg, and Tshwane. These cities, part of the C40 Cities Network,[6] have committed to achieving NZC status for all new buildings by 2030 and all existing buildings by 2050. This commitment is part of the C40 Global NZC Buildings declaration and has been supported by various international and local funding sources.

Key activities include the drafting and approval of policy documents, the creation of draft bylaws, and collaboration across multiple sectors, from private enterprises to national government bodies. Legal and financial guidance has been provided to navigate the complexities of implementation, while data collection has underpinned the evidence base for policy acceptance. Tools such as the "Getting to Zero"[7] manual and compliance software have been developed to assist the market in meeting NZC goals. Education and communication efforts are ongoing to build awareness and engagement among stakeholders.

Efforts to integrate municipal policies into national frameworks, build capacity within municipalities, provide practical guidance and financial models for NZC buildings, and facilitate public participation and community engagement have been integral to the policy development and implementation process. Much of this work has been coordinated by SEA, although there has been support from many other stakeholders, particularly in the civil society group, such as the CSIR and the GBCSA.

The above two success stories have common factors that seem to contribute to their success. In both cases, there is a lead party that leads the coordination of stakeholder engagements. Furthermore, in both cases, the coordinating party is a member of the civil society stakeholder group. These parties are well-positioned to engage across the full spectrum of the building value chain. Key success factors include regular engagements and strong personal and organisational leadership.

[6] https://www.c40.org/.

[7] https://gbcsa.org.za/wp-content/uploads/2020/08/Getting-to-Zero_2020.pdf.

6 Conclusion and Further Research

This article has examined the need for effective stakeholder engagement in construction industry in order to achieve net zero carbon building targets. It highlights that isolated NZC projects are not enough to achieve global or national decarbonisation targets, and that a coordinated industry-wide effort is required. For such an industry transition, it is crucial for all stakeholders to be aligned.

Stakeholders across the whole life cycle value chain of buildings are identified, namely, policy makers, industry actors, end users, and civil society. The role of each of these stakeholders is discussed, as well as the barriers and opportunities for implementing NZC buildings and influencing industry norms.

Ideas were collected from various stakeholder group representatives via a questionnaire. Key stakeholder activities, namely engaging with local communities and environmental groups, preparing the workforce, and education and training, were discussed with reference to each stakeholder group, mapping roles of each group in the activities. In each of these activities, certain stakeholders have a dominant role, but all have a contribution. The role of policy makers as national drivers is noted, yet policy alone cannot change the industry quickly enough to achieve net zero carbon buildings by 2050.

Two success stories are presented. These demonstrate how effective stakeholder engagement can produce change. A notable success indicator is stakeholder alignment and broad participation in activities, such as workshops and training. Factors identified that achieve alignment and action include strong leadership, clear objectives, relevant dialogue, regular engagement, and broad participation. Leadership and coordination of the industry are well-located in the civil society stakeholder group, which includes actors that cut across the whole value chain.

Acknowledgements The following individuals are acknowledged for their contributions: Marc Sherratt (March Sherratt Sustainability Architects), Rabelani Tshikalanke and Seana Nkhahle (South African Local Government Association), Peggy Mafadza (National Home Builders Regulatory Council).

References

1. Architecture 2030 (2023) About us [Online]. Available at https://www.architecture2030.org/our-mission/. Accessed 14 Aug 2024
2. Bahadorestani A, Naderpajouh N, Sadiq R (2020) Planning for sustainable stakeholder engagement based on the assessment of conflicting interests in projects. J Cleaner Prod 242
3. Bansal H, Eiselt HA (2004) Exploratory research of tourist motivations and planning. Tour Manag 25(3):387–396
4. Department for Communities and Local Government (2006) Code for sustainable homes: a step-change in sustainable home building practice. Crown, UK
5. Department for Energy Security and Net Zero (2023) Net zero government initiative: UK roadmap to net zero government emissions. Crown, UK

6. Falana J, Osei-Kyei R, Tam V (2024) Towards achieving a net zero carbon building: a review of key stakeholders and their roles in net zero carbon building whole life cycle. J Build Eng 82
7. Freeman RE (1984) Strategic management: a stakeholder approach. Pitman, London
8. Freeman RE, Phillips RA (2023) Stakeholder theory: a libertarian defense. In: R. Edward Freeman's selected works on stakeholder theory and business ethics. Springer Nature, Berlin, p 814
9. Juujärvi S, Lund P (2016) Enhancing early innovation in an urban living lab: lessons from Espoo, Finland. Technol Innov Manag Rev 6(1):17–29
10. Kogan N, Lee KJ (2014) Exploratory research on the success factors and challenges of smart city projects. Asia Pac J Inf Syst 24(2):141–189
11. Kujala J et al (2022) Stakeholder engagement: past, present, and future. Bus Soc 61(5):1136–1196
12. Lui X et al (2020) Living lab, a platform for facilitating and accelerating sustainable urban development: the case of Hiedanranta district in Tampere, Finland. Urban Plan 5(3):267–279
13. Mahmoodi M, Rasheed E, Le A (2024) Systematic review on the barriers and challenges of organisations in delivering new net zero emissions buildings. Buildings 14(6)
14. Mainardes EW, Alves H, Raposo M (2010) An exploratory research on the stakeholders of a university. J Manag Strategy 1(1):76
15. Mbaka N, Isiramen OM (2021) The changing role of an exploratory research in modern organisation. GPH Int J Bus Manag 4(12):27–36
16. Mlecnik E et al. (2020) Policy challenges for the development of energy flexibility services. Energy Policy 137
17. NDC Partnership (2024) NDC partnership [Online]. Available at https://ndcpartnership.org/country/zaf#:~:text=South%20Africa%20submitted%20its%20first,350%2D420%20MtCO2e%20by%202030. Accessed 5 Aug 2024
18. Nidhin BS, Domingo N, Bui TP, Wilkinson S (n.d.) Construction stakeholders' knowledge on zero carbon initiatives in New Zealand. Int J Build Pathol Adapt
19. Ohene E, Chan A, Darko A (2022) Review of global research advances towards net-zero emissions buildings. Energy Build 266
20. Pan W, Pan M (2020) A 'demand-supply-regulation-institution' stakeholder partnership model of delivering zero carbon buildings. Sustain Cities Soc 62
21. Passer A et al (2020) Sustainable built environment: transition towards a net zero carbon built environment. Int J Life Cycle Assess 25:1160–1167
22. Rapid Transition Alliance (2019) Back from the brink: how the world rapidly sealed a deal to save the ozone layer [Online]. Available at https://rapidtransition.org/stories/back-from-the-brink-how-the-world-rapidly-sealed-a-deal-to-save-the-ozone-layer/#:~:text=It%20took%20only%202%20years,enforcing%20the%20ban%20on%20CFCs. Accessed 26 July 2024
23. Shell and Deloitte (2023) Decarbonising construction: building a low-carbon future. Shell International B.V., Netherlands
24. Stevenson F, Kwok A (2020) Mainstreaming zero carbon: lessons for built-environment education and training. Build Cities 1(1):687–696
25. Tobi H, Kampen JK (2018) Research design: the methodology for interdisciplinary research framework. Qual Quant 52:1209–1225
26. Warburton K (2003) Deep learning and education for sustainability. Int J Sustain High Educ

Low-Carbon Construction Materials and Technologies: Innovative Technologies for Carbon Emissions Tracking in the Construction Industry

H. Mallawaarachchi, K. S. Jayakodi, and Y. J. M. Yatawatta

Abstract The construction industry is one of the major contributors to global carbon emissions due to its high material and equipment usage thus, has obtained the world's concern in implementing carbon reduction pathways. More accurate and reliable carbon emission data is one of the key elements in moving towards low-carbon emission goals of any industry. In that case, carbon emission tracking offers a huge contribution towards emissions reduction in the construction industry. Many research scholars emphasise the importance of carbon emissions tracking towards emissions reduction, but research attempts focusing on providing a comprehensive understanding of modern and innovative technologies of carbon emission tracking in the construction industry are still lacking. Hence, this research aims to review the innovative technologies for tracking carbon emissions in the construction industry through a systematic review. A Scopus-based systematic review was conducted to analyse the journal articles related to innovative technologies, carbon emissions tracking, and the construction industry and their intersections. The bibliometric techniques were used for the analysis of the results of the systematic review. Bibliometric analysis outcomes first reveal the evolution of journal articles, leading journals, leading authors, and leading countries published research on the intersection of innovative technologies, carbon emissions tracking, and the construction industry. Second, the different innovative technologies for tracking carbon emissions in the construction industry were recognised as a key contribution of the research. Further, this chapter provides a comprehensive underpinning of the conceptual development of the intersection of innovative technologies, carbon emissions tracking, and the construction industry over the years which provides a clear pathway for future research scholars in attempting to observe the practical implications of the use of innovative technologies for carbon emissions reduction in the construction industry.

Keywords Carbon emissions tracking · Construction industry · Innovative technologies · Artificial intelligence

---

H. Mallawaarachchi (✉) · K. S. Jayakodi · Y. J. M. Yatawatta
Department of Facilities Management, University of Moratuwa, Moratuwa, Sri Lanka
e-mail: hmallawarachchi@gmail.com; harshinim@uom.lk

N. Domingo et al. (eds.), *Global Net Zero Carbon Practices in Construction*, Green Energy and Technology, https://doi.org/10.1007/978-981-96-3853-6_5

1 Introduction

The construction industry holds a pivotal role in the economic and social development of any country [1, 2]. Conversely, the building construction sector can also be identified as one of the major contributors towards global carbon emissions due to its high material and equipment usage [3].

According to Jaradat et al. [2], carbon emissions from the construction industry can arise from both direct and indirect sources. Further, it described, direct emissions occur from the energy consumption for construction activities and indirect emissions are from the material extraction, production, transportation, and demolition. Rodrigo et al. [4] further classified carbon emissions into two: embodied carbon from the production stage and operational carbon from the operational stage of the buildings. The United Nations Environment Programme (UNEP) claimed that "CO_2 emissions from building and construction hit a new high, leaving the sector off track to decarbonise by 2050" [5]. Moreover, UNEP (2022) highlights that the carbon emissions from the buildings and construction sector account for 21% of global carbon emissions and are still rising.

Global warming, changes in perception patterns, crop failures, pest proliferation, sea level rises, and desertification can be identified as some examples of these consequences [6]. In addition to that, the Africa Renewable Energy Access Program (AFREA) claimed that some health issues such as lung cancer, chronic bronchitis, heart disease, and low respiratory infections are linked to carbon emissions. To mitigate all these negative consequences, the urgent necessity of moving towards a low-carbon construction industry is highlighted at both academic and industrial levels [7].

Accordingly, global attention has been paid towards the use of low-carbon construction technologies [8]. When focusing on carbon emissions tracking, many research scholars emphasise the importance of carbon emissions tracking towards emissions reduction. In addition to that, carbon emission tracking helps to improve environmental performance while reducing emissions from the construction site [9], instrumental governance parties for establishing energy conservation and emission reduction policies [10] and encourage to use of clean energy or technology and optimise the "cost, quality, schedule, and greenhouse gas (GHG) emissions" of the construction industry are some highlighted points. It is also highly considered that with the strengths blockchain has in security, transparency, and auditability, it offers a suitable alternative to traditional information systems in estimating embodied carbon in construction supply chains [4, 11]. However, a lack of research which focuses on providing a comprehensive understanding of modern and innovative technologies for carbon emissions tracking in the construction industry.

Accordingly, this research aims to review the role of innovative technologies in tracking carbon emissions in the construction industry through a systematic review. Two (02) research objectives were formulated as:

i. To review the conceptual evolution of the intersections between innovative technologies, carbon emissions tracking, and the construction industry through bibliometric analysis.
ii. To unveil the different innovative technologies of carbon emission tracking in the construction industry.

1.1 Carbon Emissions in the Construction Industry

The construction industry accounts for accountable damage to natural resources due to its higher raw material and energy consumption, waste generation, and carbon emissions [2]. Zainordin et al. [12] claimed that the construction industry holds the largest portion of global carbon emissions with rapid urbanisation. However, carbon emissions from construction can be defined as the "release of carbon into the atmosphere through human activities in the construction industry such as combustion of fossil fuel and cement production where it is stored thereby increasing the concentration of greenhouse gas emission" [12].

Carbon emissions from the construction sector can be categorised into two which are direct carbon emissions and indirect carbon emissions [13]. Further, it described direct carbon emissions occurring from the onsite activities while indirect carbon emissions occurred from the upstream and downstream supply chain, encompassing the production and transportation of material and services which is necessary for the construction activities such as manufacturing of bricks and so on. Conversely, Huang et al. [14] identified direct emissions that occurred from the onsite energy consumption and indirect emissions that occurred through the extraction of construction materials, production, transportation, demolition, and other non-building activities.

1.2 Carbon Emissions Tracking in Construction and the Use of Innovative Technologies

Carbon emission tracking is one of the main prerequisites in moving towards a low-carbon construction industry. Accordingly, many recent research studies have highlighted the urgent necessity of assessing carbon emissions in the construction industry [8]. However, Huang et al. [15] claimed that carbon emission quantification in the construction stage is complex because it includes both onsite and offsite emission activities. An extensive breakdown of the carbon emissions associated with a construction project is provided by smart construction platforms that integrate multiple digital tools [16]. Throughout the duration of a project, these platforms enable real-time monitoring, data analysis, and decision-making to maximise efficiency and reduce emissions [17]. By offering details on the carbon footprint of different building materials, sustainable materials databases play a critical role in assisting construction companies in choosing low-emission substitutes [18].

Carbon tracking in the construction sector is greatly improved by carbon accounting software, smart construction platforms, and sustainable materials databases [19]. Carbon footprint calculations for various materials and processes are provided by carbon accounting software, which carefully monitors and records emissions from construction activities [7]. This makes it possible for companies to identify the main sources of emissions and focused reduction plans into action [20]. Mésároš et al. [21] evaluate the potential of applying Artificial Intelligence (AI) for tracking the carbon footprint of the construction industry. Based on the findings, Mésároš et al. [21] claimed that the use of AI in tracking construction carbon emissions marks a significant advancement in sustainable construction.

Since innovative technologies including AI facilitate effective planning and process optimisation which reduces carbon emissions and other environmental impacts [22], this claims legitimate need of having a comprehensive analysis of the intersections of innovative technologies and carbon emissions tracking in the construction industry.

2 Research Methodology

For the present study, a systematic review of literature was used as the methodological base. Key literature published on the intersection of the fields of innovative technologies, carbon emissions tracking, and construction industry was identified through a systematic review in the Scopus database. This Scopus-based review provides initial insights to identify the innovative technologies used for carbon emissions tracking construction industry from 1991 to 2024. As the preliminary step of the review, a literature search was led in the Scopus database using the keywords of innovative technologies, carbon emissions tracking, and construction industry to search titles, abstracts, and document keywords published from 1991 to 2024 [TITLE-ABSKEY "(innovative AND technologies) AND (carbon AND emissions AND tracking) AND (construction AND industry OR construction)]".

Initially, there were 2306 articles that were further refined utilising the filters available on Scopus's search tool: (i) DOCUMENT TYPES = (Articles) AND (ii) SOURCE TYPE = (Journals) (iii) LANGUAGE = (English) to recognise high-quality literature. Then, 1662 journal articles were selected as the basis for bibliometric analysis which was used to identify papers relevant to the review. As stated by Donthu et al. [23], bibliometric analysis technique has been widely used to evaluate the contributions of research scholars in various fields, analyse publication patterns, and examine the relationships between research findings.

Hence, the analysis was conducted using two selected bibliometric indicators: the co-occurrence of words and the number of articles that intersect the fields of innovative technologies, carbon emission tracking, and the construction industry. The number of articles reflects scientific output, providing a count of the works produced by researchers. Co-occurrence, on the other hand, facilitates the recognition of the specific network of a given type of research based on its development over the years [24]. The evolution of the number of journal articles published over the years, leading journals, leading authors, and leading countries on the intersection of innovative technologies, carbon emission tracking, and the construction industry were identified through the analysis of evolving trends in journal publications over time. The literature was chosen based on the selected bibliometric indicators and reviewed to unveil the intersection between innovative technologies, carbon emission tracking, and the construction industry. Finally, the different innovative technologies for carbon emissions tracking in the construction industry were also recognised.

3 Findings and Discussion

This section presents the key research findings related to two major areas: (i) Outcomes of bibliometric analysis and (ii) Innovative technologies for carbon emissions tracking construction industry.

3.1 *Outcomes of the Bibliometric Analysis*

As the initial step, the data derived from the Scopus-based systematic review were analysed using bibliometric analysis to identify innovative technologies for carbon emission tracking in the construction industry. The results, covering the period from 1991 to 2024, are organised under four key headings: (i) Evolution of the number of journal articles, (ii) Leading journals, (iii) Leading authors in the field, and (iv) Leading countries published more on innovative technologies and carbon emissions tracking research in construction industry.

3.1.1 Evolution of the Number of Journal Articles Published on Innovative Technologies and Carbon Emissions Tracking in Construction Industry

In the systematic review, 1662 journal articles published between 1991 and the first 07 months of 2024 were selected for analysis. The evolution of the number of journal articles published on identifying innovative technologies for carbon emission tracking in the construction industry in the Scopus database is presented in Fig. 1.

Upon analysis, it was found that the total number of journal articles published on the intersection of innovative technologies, carbon emission tracking, and the construction industry in Scopus was only 1, which was published in 1991. No publications were identified from 1992 to 1998. There was only 01 article published in 2003, 2008, and 2009, with no journal publications between 2000 and 2009. However,

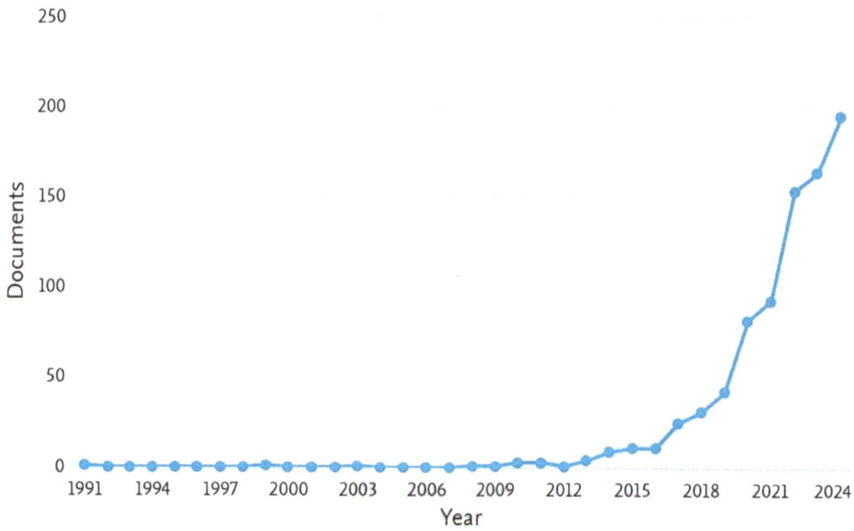

Fig. 1 Evolution of the number of journal articles

the number increased to 03 articles in both 2010 and 2011. In 2012 and 2013, 01 and 04 articles were published, respectively.

Subsequently, there was a significant rise to 09 articles in 2014, 11 in 2015, and 11 again in 2016. The numbers continued to grow, with 25 articles published in 2017, 31 in 2018, 42 in 2019, 82 in 2020, 93 in 2021, 155 in 2022, and 164 in 2023. In the first 07 months of 2024, 191 journal articles had already been published.

Overall, the graph illustrates a consistent growth in publications on the intersection of innovative technologies, carbon emission tracking, and the construction industry, from one article in 1991 to a peak of 191 articles in the first 07 months of 2024. However, many of the published articles focused on analysing the use of innovative technologies to reduce energy costs and identify the challenges and strategies for carbon emission tracking and carbon footprint management in the construction industry. Notably, a systematic review study conducted by Fang et al. [25] mentioned the building and infrastructure carbon emission management-related studies to identify the information necessary to support the project's critical decision-making. Secondly, the study identified the critical decision-making supporting information and management steps and built a generalised carbon footprint management adoption decision-making framework to support carbon project leaders' important decision-making.

And also, Sizirici et al. [26] identified range of latest techniques for reducing the carbon footprint from construction stage to operation stage including using alternative additives in building materials, enhancing design, recycling construction waste, promoting alternative water resources, increasing water technology efficiency, and developing novel systems to improve the sustainability of the construction industry. Despite this growing body of research, no articles were found that specifically addressed the intersection of innovative technologies, carbon emission tracking, and the construction industry. Therefore, this study focuses on identifying innovative technologies for carbon emission tracking in the construction industry.

3.1.2 Leading Journals Published on Innovative Technologies and Carbon Emissions Tracking in Construction Industry

The systematic review identified the leading journals that have published the most articles on the intersection of innovative technologies, carbon emission tracking, and the construction industry from 1991 to 2024. "Sustainability Switzerland" was the leading journal, publishing the highest number of articles (55 articles) on this topic during the review period. The "Journal of Cleaner Production" was the second leading journal, with 42 articles. Other significant journals include "Energy" (20 articles) and "Environmental Science and Pollution Research" (15 articles), which have also focused on publishing research at the intersection of innovative technologies, carbon emission tracking, and the construction industry from 1991 to 2024. All articles have started their publications from 2016.

The fluctuations in leading journal publications from 1991 to 2024 are shown in Fig. 2. In 2016 and 2017, the leading journal for the intersection of innovative technologies, carbon emission tracking, and the construction industry was "Sustainability Switzerland", with 01 publication each year. This number increased to 02 in 2018 and to 08 in 2019. However, in 2020 and 2021, the number of publications dropped to 04 articles. In 2022, it peaked at more than 15 articles, but in 2023 and 2024, it slightly declined to 07 articles each year.

The "Journal of Cleaner Production" began publishing on this topic in 2017 with 06 articles, showing good progress compared to other journals. However, in 2018,

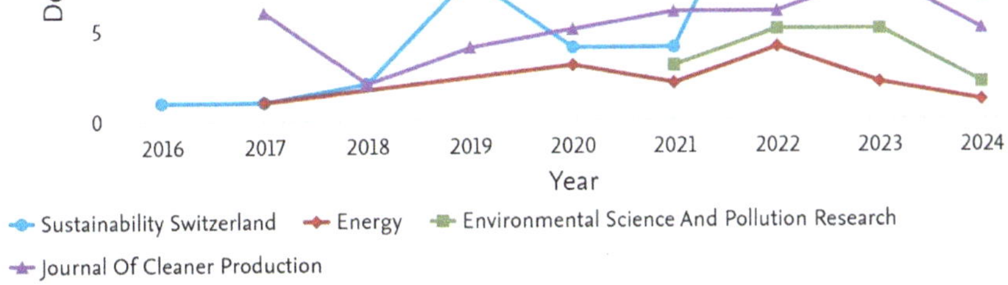

Fig. 2 Leading journals

the number dropped to 03 articles. After 2018, the number of publications gradually increased, reaching 07 articles in 2023. In the first seven months of 2024, it experienced a slight decline to 05 articles. The "Energy" journal started publishing on this intersection in 2017, with a gradual increase until 2020, when 03 articles were published. After 2020, there was a decline, but in 2021 and 2022, it peaked at 04 articles. By 2024, the number slightly dropped to 1 article. The "Journal of Environmental Science and Pollution Research" began publishing on this topic in 2021 with 03 articles. In 2022 and 2023, it published 5 articles each year, and in the first seven months of 2024, it published 01 article.

3.1.3 Leading Authors Published on Innovative Technologies and Carbon Emissions Tracking in the Construction Industry

Figure 3 presents the leading authors who have published the most papers on the intersection of innovative technologies, carbon emission tracking, and the construction industry from 1991 to 2024.

According to the analysis, Tang B. Z. is the most prolific author, with 10 records in Scopus during this period. The second-highest number of records is held by Heidari A., with 05 articles. Dong L. and Lam J. W. Y. each have published 4 articles, placing them third. All other authors have published 03 articles each. Due to the relatively limited focus on the intersection of innovative technologies, carbon emission tracking, and the construction industry, the number of articles published by individual authors is comparatively low.

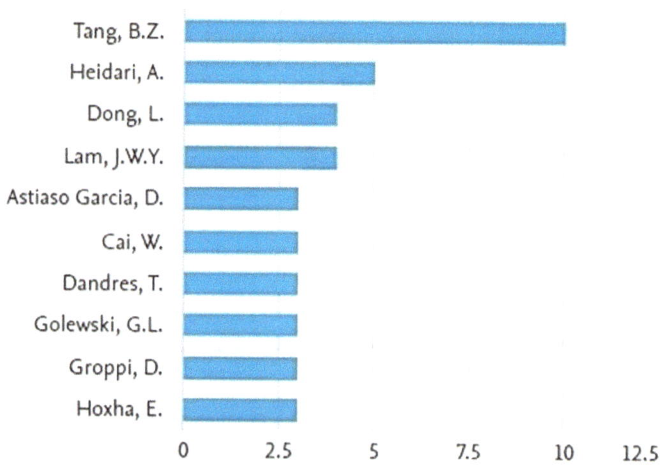

Fig. 3 Leading authors

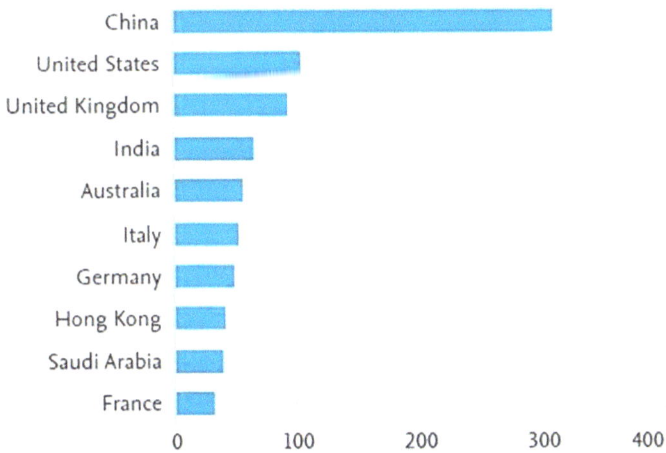

Fig. 4 Leading countries

3.1.4 Leading Countries Engaged in Innovative Technologies and Carbon Emission Tracking Research

Figure 4 highlights the number of countries focusing on innovative technology and carbon emission tracking research. China leads with over 300 publications, establishing itself as a top country in innovative technology research. The United States follows with more than 100 articles, and the United Kingdom is close behind with almost 100 articles. Additionally, India, Australia, and Italy have each published more than 50 articles on this area.

3.2 Innovative Technologies for Carbon Emissions Tracking in the Construction Industry

The construction industry is increasingly focusing on innovative technologies to track and mitigate carbon emissions [27]. The application of machine learning and Artificial Intelligence (AI) to the optimisation of renewable energy systems is one significant development [28]. Through precise energy consumption prediction and optimal building energy use, these technologies improve energy efficiency and lower emissions [28]. For example, highly accurate electricity usage forecasts made possible by AI models have made it possible to more effectively incorporate renewable energy sources, such as solar and wind power, into building projects [29]. Further, Mésároš et al. [21] discussed several artificial intelligence integrated tools and technologies for tracking construction carbon emissions such as the CarbonCure application, SimScale cloud-based simulation platform, ClimateClever application, Bentley iTwin platform, and so on. Moreover, Giannelos et al. [30] mentioned that large datasets are analysed by AI and machine learning to predict emissions and optimise construction processes [30]. By simulating the effects of various building techniques and materials on carbon emissions, these technologies facilitate more informed decision-making [31]. One of the modern technologies used in the construction sector is Building Information Modelling (BIM) [32]. Construction projects benefit from the integration of BIM, which provides comprehensive information on materials, processes, and energy consumption [32].

A BIM-based model was developed by Marzouk et al. [33] for tracking both direct and indirect carbon emissions in the construction industry. Accordingly, the proposed model mainly consists of two sources of external inputs as BIM model and the Athena Impact Estimator. The BIM model is created by using Autodesk Revit and it acts as a database while Athena Impact Estimator calculates the environmental performance of the construction project including carbon footprint, particulate matter, ozone-depleting particles, and so on. Further, Hussein and Naimi [34] focused on implementing a 7D BIM Modelling into construction project management to enhance and monitor sustainability principles by using a case study in Iraq. In this research, 7D BIM technology is used for different purposes including carbon and GHG emissions mitigation, sustainability management, and adverse environmental impact alleviation. In order to reduce carbon emissions, this makes precise lifecycle analysis and design optimisation possible [35]. BIM greatly improves carbon tracking in the construction industry by integrating and visualising data about materials, energy use, and emissions.

By enabling the inclusion of material databases with corresponding carbon footprints, BIM helps Lifecycle Assessment (LCA) and encourages the use of low-carbon materials [36]. Furthermore, BIM makes it possible to quantify the environmental impact of a building at every stage of its lifecycle, from design and construction to operation and demolition [37]. By optimising designs and identifying high-emission areas, this comprehensive data management and visualisation capability helps lower the overall carbon footprint of construction projects [7]. Zhang and Zhang [38] developed an automated carbon planning, monitoring, and forecasting system for construction projects integrating the Building Information Modelling (BIM) and Earned Value Method (EVM).

Initially, a carbon-loaded planning technique is developed utilising a building element-level carbon database in this research project. Then, this is followed by the creation of an advanced EVM model, which is integrated into the system to enable tracking and forecasting of carbon emissions in construction. Subsequently, the carbon-loaded planning technique and EVM model are integrated and developed to enable automated data exchange and visualisation. Finally, a case study was carried out to demonstrate and validate its functionality. Blockchain technology is emerging as a promising tool for accurate carbon emissions tracking which ensures traceability and transparency in construction supply chains [4, 39]. It offers an immutable ledger that documents every transaction, ensuring that all building materials are certified and sourced ethically [39]. Blockchain reduces the risk of using non-compliant materials by tracking the entire supply chain process, from the extraction of raw materials to the finished product [40]. This ensures transparency and accountability at every stage of the process. Accordingly, the construction industry would dramatically improve the accuracy and reliability of tracking carbon emissions and reduce overall carbon footprints [4, 11]. Furthermore, construction companies can purchase and sell carbon credits as needed to offset their emissions through blockchain's transparent and secure carbon credit trading [41]. Companies can promote compliance with environmental regulations by tracking their carbon footprint in real time and modifying their operations to stay within their carbon budget [42].

Further, Yevu et al. [19] introduced a digital-twin-enabled prefabricated supply chain for carbon emission tracking in construction projects. Accordingly, Radio Frequency Identification (RFID), Global Positioning systems (GPS), and laser scanners and sensors have been deployed for real-time data acquisition in this real-time emission monitoring system. Further, this research revealed several future research directions to enhance construction processes and emissions monitoring. Moreover, with the highest significance of construction carbon emissions tracking, Liu et al. [43] proposed a cyber-physical system integrated real-time carbon emission monitoring system for prefabricated construction. Accordingly, this system facilitates real-time carbon emission tracking from production to installation of prefabricated components. The system comprises with main three layers as sensing layer, computing, and interaction layer. Accordingly, it integrates with acceleration sensors, barometric sensors, and GPA sensors in the sensing layer for data collection in real time; a remote server and a database in the computing layer for data process while desktop and mobile devices in the interaction layer for data presentation. Further, General Packet Radio Service (GPRS) and wireless local area networks (WLANs) are used to media for communication throughout the system [43].

Internet of Things (IoT) sensors are essential for tracking carbon emissions in the construction sector because they offer real-time data and insights that facilitate more efficient monitoring and management of carbon emissions [44]. The sensors are designed to continuously monitor multiple parameters, including energy consumption, machinery usage, and emissions levels on construction sites [45]. This enables prompt corrective actions by enabling the immediate detection of inefficiencies and high-emission sources. Construction managers can optimise usage and maintenance schedules to minimise emissions by using sensors attached to construction equipment and machinery to track fuel consumption and operational efficiency [46]. IoT sensors can also keep track on site environmental parameters like humidity and temperature to make sure materials are used and stored in the best possible ways, reducing on waste and related emissions [47]. In addition to that, an IoT-based GHG Emission Monitoring (GEM) system was proposed by Deng et al. [10] for real-time emissions monitoring in prefabricated components manufacturing.

Here, Radio Frequency Identification (RFID) sensors and laser sensors are used for data acquisition and a data service platform is used for data transmission and visualisation. As a result, this system facilitates to management of irregular GHG emissions, possible emission risk identification, and explores emission reduction strategies in the construction sector [10]. Smart grid integration maximises the use of renewable energy sources and improves energy management, and it is essential for carbon tracking in the construction industry [48]. Real-time monitoring and control of energy consumption is made possible by smart grids, which allow two-way communication between the energy grid and the construction site. By transferring usage to periods when renewable energy is available and grid demand is low, this integration helps construction sites draw energy more efficiently, decreasing their dependency on fossil fuels and their carbon emissions [49]. Furthermore, smart grids make it easier to incorporate onsite renewable energy sources, like solar or wind turbines [49]. By ensuring that the cleanest energy is used possible, construction activities can further reduce their carbon footprint [50].

Additionally, construction managers can identify inefficiencies and put strategies for energy conservation and emission reduction into action with the help of smart grids, which offer detailed data on patterns of energy usage and carbon emissions [51]. Smart grids can forecast energy needs and optimise energy distribution through the use of predictive modelling and advanced analytics [48]. This allows for the most environmentally and energy-efficient construction projects [50]. In addition to these research attempts, Heydarian and Golparvar-Fard [52] has proposed a visual monitoring framework for integrated productivity and carbon footprint control of construction operations; Heydarian et al. [53] have developed an automated carbon emission benchmarking and monitoring system for earthmoving operations in the construction industry; Memarzadeh and Golparvar-Fard [54] created a construction-embodied carbon footprint monitoring and visualisation model using DnAR—N-dimensional Augmented Reality.

As unveiled above, low-carbon construction can be served by various innovative technologies significantly. However, the practical adaptation of them must be thoroughly examined in order to utilise the technologies more efficiently in order to achieve zero carbon goals of the construction industry.

4 Conclusion and Further Research

The construction industry is placing a growing emphasis on the utilisation of innovative technologies to monitor and reduce carbon emissions. Among the other various technologies, the application of machine learning and AI to the optimisation has obtained a world attention. Hence, this research focused on reviewing the existing literature published during the period of 1991–2024. The literature was selected based on the selected bibliometric indicators and studied to reveal the intersection between innovative technologies, carbon emission tracking, and the construction industry and to review the different innovative technologies for carbon emissions tracking in the construction industry.

As revealed through bibliometric analysis, there is a consistent growth in publications on the intersection of innovative technologies, carbon emissions tracking, and the construction industry during the period from 1991 to 2024. However, many of the published articles focused on analysing the use of innovative technologies to reduce energy costs and identify the challenges and strategies for carbon emission tracking and carbon footprint management in the construction industry. Sustainability

Switzerland and the Journal of Cleaner Production were leading journals where Tang and Heidari are the leading authors who have published more on the intersection of innovative technologies, carbon emission tracking, and the construction industry. The selected articles were then reviewed and various innovative technologies, such as machine learning and AI, blockchain, IoT, BIM, digital twin, and smart grid were revealed that can be used to track carbon emissions in construction industry.

This chapter offers a comprehensive basis for understanding the conceptual evolution of how innovative technologies intersect with carbon emissions tracking in the construction industry. It outlines the historical progression and provides a clear direction for future research, guiding scholars in exploring the practical applications of these technologies for reducing carbon emissions within the industry. Since this research is limited to understand the conceptual development of the intersection between innovative technologies, carbon emissions tracking, and construction industry, the next stages of this research aimed at investigating the application of each technology alone in carbon emissions tracking, enhancing the applicability of innovating technologies for developed and developing countries in achieving carbon reduction goals and developing a AI-based carbon emissions tracking model for construction industry.

References

1. Huang L, Krigsvoll G, Johansen F, Liu Y, Zhang X (2018) Carbon emission of global construction sector. Renew Sustain Energy Rev 81:1906–1916. Elsevier Ltd. https://doi.org/10.1016/j.rser.2017.06.001
2. Jaradat H, Alshboul OAM, Obeidat IM, Zoubi MK (2024) Green building, carbon emission, and environmental sustainability of construction industry in Jordan: awareness, actions and barriers. Ain Shams Eng J 15(2). https://doi.org/10.1016/j.asej.2023.102441
3. Labaran YH, Mathur VS, Farouq MM (2021) The carbon footprint of construction industry: a review of direct and indirect emission. J Sustain Constr Mater Technol 6(3):101–115
4. Rodrigo MNN, Perera S, Senaratne S, Jin X (2020) Potential application of blockchain technology for embodied carbon estimating in construction supply chains. Buildings 10(8):140. https://doi.org/10.3390/BUILDINGS10080140
5. De la Colina Martínez AL, Delgado Hernández DJ (2024) Circular cement decarbonisation: towards a net-zero built environment. In: Environmental engineering and waste management: recent trends and perspectives. Cham, Springer Nature Switzerland, (pp 269-296)
6. Farooq MU, Shahzad U, Sarwar S, Zaijun L (2019) The impact of carbon emission and forest activities on health outcomes: empirical evidence from China. Environ Sci Pollut Res 26(13):12894–12906. https://doi.org/10.1007/s11356-019-04779-x
7. Wong JKW et al (2013) Toward low-carbon construction processes: the visualisation of predicted emission via virtual prototyping technology. Autom Constr 33:72–78. Available at https://doi.org/10.1016/j.autcon.2012.09.014
8. Lu H, You K, Feng W, Zhou N, Fridley D, Price L (2024) Reducing China's building material embodied emissions: opportunities and challenges to achieve carbon neutrality in building materials. Iscience 27(3)
9. Seo J, Park J, Oh Y, Park S (2016) Estimation of total transport CO2 emissions generated by medium-and heavy-duty vehicles (MHDVs) in a sector of Korea. Energies 9(8):638

10. Deng T, Huang Y, Li Z, Wang N, Wang S, Zou Y, ... Fan S (2018) Numerical simulations for the sources apportionment and control strategies of PM2. 5 over Pearl River Delta, China, part II: vertical distribution and emission reduction strategies. Sci Total Environ 634:1645–1656
11. Rodrigo MNN, Perera S, Senaratne S, Jin X (2022) Systematic development of a data model for the blockchain-based embodied carbon (BEC) estimator for construction. Eng Constr Archit Manag 29(9):3311–3330. https://doi.org/10.1108/ECAM-02-2021-0130/FULL/XML
12. Zainordin N, Zahra DBF (2020) Factors contributing to carbon emission in construction activity. In: Third international conference on separation technology 2020 (ICoST 2020). Atlantis Press, pp 176–182
13. Kokoni S, Skea J (2014) Input–output and life-cycle emissions accounting: applications in the real world. Clim Policy 14(3):372–396
14. Huang W, Li, F, Cui SH, Huang L, Lin JY (2017) Carbon footprint and carbon emission reduction of urban buildings: a case in Xiamen City, China. Procedia Eng 198:1007–1017. https://doi.org/10.1016/j.proeng.2017.07.146
15. Huang Z, Zhou H, Mia Z, Tang H, Lin B, Zhuang W (2024) Life-cycle carbon emissions (LCCE) of buildings: implications, calculations, and reductions. Engr 35:115–139. Elsevier Ltd. https://doi.org/10.1016/j.eng.2023.08.019
16. Yan J et al (2022) Digital tools for revealing and reducing carbon footprint in infrastructure, building, and city scopes. Buildings 12(8):1097. Available at https://doi.org/10.3390/buildings12081097
17. Rao AS et al (2022) Real-time monitoring of construction sites: sensors, methods, and applications. Autom Constr 136:104099. Available at https://doi.org/10.1016/j.autcon.2021.104099
18. Chen Y, Ng ST, Hossain MU (2018) Approach to establish carbon emission benchmarking for construction materials. Carbon Manag 9(6):587–604. Available at https://doi.org/10.1080/17583004.2018.1522094
19. Yevu SK et al (2023) Digital twin-enabled prefabrication supply chain for smart construction and carbon emissions evaluation in building projects. J Build Eng 78:107598. Available at https://doi.org/10.1016/j.jobe.2023.107598
20. Ozawa-Meida L et al (2013) Measuring carbon performance in a UK university through a consumption-based carbon footprint: De Montfort University case study. J Cleaner Prod 56:185–198. Available at https://doi.org/10.1016/j.jclepro.2011.09.028
21. Mésároš P, Smetanková J, Behúnová A, Krajníková K (2024) The potential of using artificial intelligence (AI) to analyse the impact of construction industry on the carbon footprint. Mobile Netw Appl. https://doi.org/10.1007/s11036-024-02368-y
22. Adewale BA, Ene VO, Ogunbayo BF, Aigbavboa CO (2024) A systematic review of the applications of AI in a sustainable building's lifecycle. Buildings 14(7):2137. https://doi.org/10.3390/buildings14072137
23. Donthu N, Kumar S, Mukherjee D, Pandey N, Lim WM (2021) How to conduct a bibliometric analysis: an overview and guidelines. J Bus Res 133:285–296. https://doi.org/10.1016/j.jbusres.2021.04.070
24. Mokhtarpour R, Khasseh AA (2021) Twenty-six years of LIS research focus and hot spots, 1990–2016: a co-word analysis. J Inf Sci 47(6):794–808. https://doi.org/10.1177/0165551520932119
25. Fang Z et al. (2023) A systematic literature review of carbon footprint decision-making approaches for infrastructure and building projects. Appl Energy 335:120768. Available at https://doi.org/10.1016/j.apenergy.2023.120768
26. Sizirici B et al (2021) A review of carbon footprint reduction in construction industry, from design to operation. Materials 14(20):6094. Available at https://doi.org/10.3390/ma14206094
27. Huisingh D et al (2015) Recent advances in carbon emissions reduction: policies, technologies, monitoring, assessment and modelling. J Cleaner Prod 103:1–12. Available at https://doi.org/10.1016/j.jclepro.2015.04.098
28. Entezari A et al (2023) Artificial intelligence and machine learning in energy systems: a bibliographic perspective. Energy Strategy Rev 45:101017. Available at https://doi.org/10.1016/j.esr.2022.101017

29. Boza P, Evgeniou T (2021) Artificial intelligence to support the integration of variable renewable energy sources to the power system. Appl Energy 290:116754. Available at https://doi.org/10.1016/j.apenergy.2021.116754
30. Giannelos S et al (2024) Machine learning approaches for predictions of CO_2 emissions in the building sector. Electric Power Syst Res 235:110735. Available at https://doi.org/10.1016/j.epsr.2024.110735
31. Si J et al (2016) Assessment of building-integrated green technologies: a review and case study on applications of multi-criteria decision making (MCDM) method. Sustain Cities Soc 27:106–115. Available at https://doi.org/10.1016/j.scs.2016.06.013
32. Ahmed S (2018) Barriers to implementation of building information modeling (BIM) to the construction industry: a review. J Civ Eng Constr 7(2):107–113. Available at https://doi.org/10.32732/jcec.2018.7.2.107
33. Marzouk M, Abdelkader EM, Al-Gahtani K (2017) Building information modeling-based model for calculating direct and indirect emissions in construction projects. J Clean Prod 152:351–363. https://doi.org/10.1016/j.jclepro.2017.03.138
34. Hussein HM, Naimi S (2024) Implementing 7D BIM modeling into construction project management to enhance and monitor sustainability principles: a case study in Iraq. J Ecol Eng 2024(2):39–48
35. Hao JL et al (2020) Carbon emission reduction in prefabrication construction during materialization stage: a BIM-based life-cycle assessment approach. Sci Total Environ 723:137870. Available at https://doi.org/10.1016/j.scitotenv.2020.137870
36. Xu J et al (2023) Integrating IoT and BIM for tracking and visualising embodied carbon of prefabricated buildings. Build Environ 242:110492. Available at https://doi.org/10.1016/j.buildenv.2023.110492
37. Jalaei F, Zoghi M, Khoshand A (2021) Life cycle environmental impact assessment to manage and optimize construction waste using building information modeling (BIM). Int J Constr Manag 21(8):784–801. Available at https://doi.org/10.1080/15623599.2019.1583850
38. Zhang X, Zhang X (2023) An automated project carbon planning, monitoring and forecasting system integrating building information model and earned value method. J Clean Prod 397:136526. https://doi.org/10.1016/j.jclepro.2023.136526
39. Nawari N, Ravindran S (2019) Blockchain and building information modeling (BIM): review and applications in post-disaster recovery. Buildings 9(6):149. Available at https://doi.org/10.3390/buildings9060149
40. Nikolakis W, John L, Krishnan H (2018) How blockchain can shape sustainable global value chains: an evidence, verifiability, and enforceability (EVE) framework. Sustainability 10(11):3926. Available at https://doi.org/10.3390/su10113926
41. Blumberg G, Sibilla M (2023) A carbon accounting and trading platform for the UK construction industry. Energies 16(4):1566. Available at https://doi.org/10.3390/en16041566
42. Al Sadawi A et al (2021) A comprehensive hierarchical blockchain system for carbon emission trading utilizing blockchain of things and smart contract. Technol Forecast Soc Change 173:121124. Available at https://doi.org/10.1016/j.techfore.2021.121124
43. Liu Z, Ciais P, Deng Z, Davis SJ, Zheng B, Wang Y, Cui D, Zhu B, Dou X, Ke P, Sun T, Guo R, Zhong H, Boucher O, Bréon FM, Lu C, Guo R, Xue J, Boucher E, … Chevallier F (2020) Carbon monitor, a near-real-time 95 daily dataset of global CO2 emission from fossil fuel and cement production. Sci Data 7(1). https://doi.org/10.1038/s41597-020-00708-7
44. Lai KE et al (2023) Quantification process of carbon emissions in the construction industry. Energy Build 289:113025. Available at https://doi.org/10.1016/j.enbuild.2023.113025
45. Shahnavaz F, Akhavian R (2022) Automated estimation of construction equipment emission using inertial sensors and machine learning models. Sustainability 14(5):2750. Available at https://doi.org/10.3390/su14052750
46. Masih-Tehrani M, Ebrahimi-Nejad S, Dahmardeh M (2020) Combined fuel consumption and emission optimization model for heavy construction equipment. Autom Constr 110:103007. Available at https://doi.org/10.1016/j.autcon.2019.103007

47. Ramírez-Moreno MA et al (2021) Sensors for sustainable smart cities: a review. Appl Sci 11(17):8198. Available at https://doi.org/10.3390/app11178198
48. Rathor SK, Saxena D (2020) Energy management system for smart grid: an overview and key issues. Int J Energy Res 44(6):4067–4109. Available at https://doi.org/10.1002/er.4883
49. Kolokotsa D (2016) The role of smart grids in the building sector. Energy Build 116:703–708. Available at https://doi.org/10.1016/j.enbuild.2015.12.033
50. Oró E et al (2015) Energy efficiency and renewable energy integration in data centres: strategies and modelling review. Renew Sustain Energy Rev 42:429–445. Available at https://doi.org/10.1016/j.rser.2014.10.035
51. Elkhorchani H, Grayaa K (2016) Novel home energy management system using wireless communication technologies for carbon emission reduction within a smart grid. J Cleaner Prod 135:950–962. Available at https://doi.org/10.1016/j.jclepro.2016.06.179
52. Heydarian A, Golparvar-Fard M (2011) A visual monitoring framework for integrated productivity and carbon footprint control of construction operations
53. Heydarian A, Memarzadeh M, Golparvar-Fard M, Students G (2012) Automated benchmarking and monitoring of earthmoving operation's carbon footprint using video cameras and a greenhouse gas estimation model
54. Memarzadeh M, Golparvar-Fard M (2012) Monitoring and visualization of building construction embodied carbon footprint using DnAR—N-dimensional augmented reality models

Design for Low Carbon in the Construction Industry

M. Samaratunga and S. Li

Abstract The construction industry is a significant source of global carbon emissions, with both operational carbon (emissions from building energy use) and embodied carbon (emissions from materials and construction processes) contributing to the sector's environmental footprint. As the demand for sustainable development intensifies, there is an urgent need to reduce carbon emissions in construction to meet climate goals. This chapter explores strategies to lower operational carbon through energy optimisation and reduce embodied carbon through the use of low-carbon materials and sustainable construction methods. The rationale behind this research is to provide a clear understanding of how innovative design and material choices can drive the industry towards a low-carbon future. This chapter is based on a critical literature review, which involves a comprehensive analysis of current research on energy-efficient building design and the development of low-carbon materials. By examining a wide range of studies, this chapter synthesises key insights on passive and active design strategies for energy optimisation, material innovations and construction methods that contribute to carbon reduction. Additionally, the review includes the use of assessment tools such as Life Cycle Assessment (LCA) and Environmental Product Declarations (EPDs), which provide insights into the carbon impacts of different design decisions and materials. The findings from the literature review indicate that both passive design strategies (such as building orientation, natural ventilation and daylighting) and active systems (such as energy-efficient HVAC and renewable energy integration) are effective in reducing operational carbon. The use of Building Information Modelling (BIM) and digital tools further supports energy optimisation by allowing for more accurate simulations of building performance. In terms of embodied carbon, emerging low-carbon materials like bio-based and recycled products, along with sustainable construction methods like modular construction, have demonstrated significant potential to reduce emissions. These findings can be highly useful in addressing gaps in the construction industry's approach to carbon

M. Samaratunga (✉)
Centre for Smart Modern Construction, School of Engineering, Design and Built Environment, Western Sydney University, Sydney, Australia
e-mail: marini.sam@westernsydney.edu.au

S. Li
School of Built Environment, Massey University, Auckland, New Zealand

N. Domingo et al. (eds.), *Global Net Zero Carbon Practices in Construction*, Green Energy and Technology, https://doi.org/10.1007/978-981-96-3853-6_6

reduction. By providing a comprehensive review of current strategies, this study can guide building designers, architects, engineers and policymakers in adopting proven methods and materials to reduce both operational and embodied carbon. The insights can also inform sustainability standards, help develop new regulations for low-carbon construction and support community efforts in building more eco-friendly infrastructure. Ultimately, this chapter contributes to the body of knowledge necessary to push the construction industry towards achieving carbon neutrality.

1 Introduction

The building sector consumes up to 40% of all energy and contributes up to 30% of global annual carbon emissions [88]. Carbon emissions in buildings are twofold: operational carbon emissions and embodied carbon emissions. In Australia, data from the Department of Climate Change, Energy, the Environment and Water shows buildings account for around 19% of total operational energy use and 18% of direct carbon emissions [12]. Historically, embodied emissions have received far less attention. However, there has been a recent shift towards focusing on embodied carbon, driven by the realisation that it accounts for a significant portion of a building's total life cycle greenhouse gas (GHG) emissions, sometimes exceeding 50% [31]. As a result, there is growing recognition of the need to address embodied carbon alongside operational carbon to achieve meaningful reductions in overall emissions.

The environmental impact of the construction industry is profound, affecting both developed and developing nations. As urbanisation accelerates, the demand for new infrastructure and buildings increases, leading to higher carbon emissions. Operational carbon, the emissions generated by a building during its use, can be mitigated by improving energy efficiency [61]. However, embodied carbon, which results from the extraction, processing and transportation of materials, as well as construction processes, continues to be a challenge [19, 64]. Despite awareness of the issue, efforts to address embodied carbon have been limited, with most attention focused on operational emissions.

In many regions, such as Europe and North America, policies have been enacted to reduce operational carbon emissions, yet similar regulatory frameworks for embodied carbon are lagging [82]. Australia and New Zealand are beginning to address this issue with initiatives like the inclusion of a materials index in the Building Sustainability Index (BASIX) in New South Wales, which encourages the use of low-carbon materials, and the development of an embodied carbon indicator by the National Australian Built Environment Rating System (NABERS) [30, 98]. Despite these efforts, significant challenges remain, including the lack of uniform methods for calculating embodied carbon, which can hinder consistent and effective regulation [16, 29]. This disparity highlights a crucial knowledge gap, particularly in the integration of low-carbon materials and construction methods. Bridging this gap requires a comprehensive approach that addresses both operational and embodied carbon in building design. In addition to carbon assessing methods, there is a need to investigate how sustainable materials and low-carbon construction methods can be incorporated into building design.

Therefore, this chapter aims to explore and propose design strategies that effectively reduce both operational and embodied carbon in building projects. The specific objectives are to identify design strategies for minimising operational carbon through energy optimisation, explore approaches for lowering embodied carbon via sustainable construction methods and materials and investigate how existing carbon assessment tools can influence the use of innovative materials and technologies in supporting low-carbon design. Together, these objectives contribute to a comprehensive approach for enhancing sustainability in the construction industry.

A comprehensive literature review is deployed in this chapter to achieve the specific aim and objectives, which include exploring design strategies for energy optimisation to reduce operational carbon, exploring sustainable construction methods and materials for lowering embodied carbon and investigating on the existing carbon assessment tools can influence the use of innovative materials and technologies in supporting low-carbon design. This approach allows for a thorough analysis of existing research and industry practices, providing a foundation for identifying best practices and recommending strategies that address the chapter's objectives.

The findings emphasise the importance of integrating both passive and active design strategies to reduce operational carbon, including optimising building orientation, using renewable energy systems and implementing smart technologies for energy efficiency [101]. The chapter also highlights the need to select low-carbon materials early in the design phase, as embodied carbon can account for up to 50–75% of a building's total life cycle emissions [22,37]. Sustainable construction practices, such as modular construction and innovative materials like bio-based products and recycled composites, can significantly lower emissions [92]. The literature directs attention to using current carbon assessment tools to inform design decisions, such as Life Cycle Assessment (LCA) for evaluating embodied and operational carbon, Environmental Product Declarations (EPDs) for product-specific impact data, Whole-Building Carbon Assessments and digital tools for carbon modelling [28]. This comprehensive approach ensures that both operational and embodied carbon are effectively addressed throughout the building life cycle, supporting the development of low-carbon buildings.

By addressing both operational and embodied carbon, this chapter contributes to the ongoing efforts to decarbonise the construction industry and supports the broader goal of achieving carbon-neutral buildings. This work is particularly relevant as the world moves towards more stringent climate targets, such as those outlined in the Paris Agreement. In particular, the chapter provides practical guidance for integrating sustainable materials and energy-efficient design into modern construction projects, contributing to the global effort to achieve carbon neutrality in the building sector. It intends to influence architects, engineers, building developers and policymakers who can use the discussion to make more informed decisions on reducing carbon emissions.

2 Reducing Operational Carbon Through Energy-Efficient Building Design

2.1 Integrating Energy Efficiency in Building Design

Energy efficiency is a fundamental component of sustainable building design, aimed at reducing operational carbon emissions over a building's life cycle. Building designs through intelligent design choices has a significant role in minimising operational carbon, which refers to emissions from the energy consumed during building use [63, 73]. Key design principles for energy efficiency throughout the building life cycle involve a holistic approach. This encompasses not only the initial design and construction phases but also how buildings will operate and be maintained over time. Life cycle energy analysis, which evaluates both embodied and operational energy, enables more comprehensive planning [23]. For example, in Australia the National Construction Code (NCC) mandates minimum 7-star energy ratings for new residential buildings, leading to an average 25% reduction in household energy consumption [11]. Similarly, in New Zealand, the Building Code has been progressively updated to improve energy efficiency standards for residential and commercial buildings, focusing on insulation, ventilation and heating systems [67]. These regulations emphasise the importance of reducing operational energy to align with long-term sustainability targets. By integrating life cycle energy considerations, policymakers can ensure that both the construction process and building operation contribute to broader climate change mitigation goals.

A holistic approach to operational energy optimisation in building design encompasses various principles that integrate environmental, social and technological considerations throughout the building life cycle [47, 91]. This approach not only enhances energy performance but also ensures occupant comfort and well-being. There are several building design strategies and approaches that focus on enhancing energy efficiency. Figure 1 indicates some of those key design strategies: smart building technologies, energy-efficient lighting, energy-efficient HVAC system, energy systems integration and renewable energy, sustainable materials and construction methods, insulation and airtightness, and passive solar design and climatic design [42, 60, 93]. These design strategies aim to reduce energy consumption while maintaining or improving occupant comfort and operational performance.

Employing smart building technologies to monitor and control energy use, designing for future adaptability [60] and selecting energy-efficient lighting and HVAC systems can lead to significant reductions in energy consumption throughout the building's operational life [42]. Integrated energy systems with renewable energy sources such as PV systems into building elements [44, 93] is an innovative and holistic design approach, which will be elaborated in Sect. 2.3. Appropriate building material selection and energy-efficient characteristics of building materials can reduce the emission of greenhouse gases and reduce the energy consumption [4, 81].

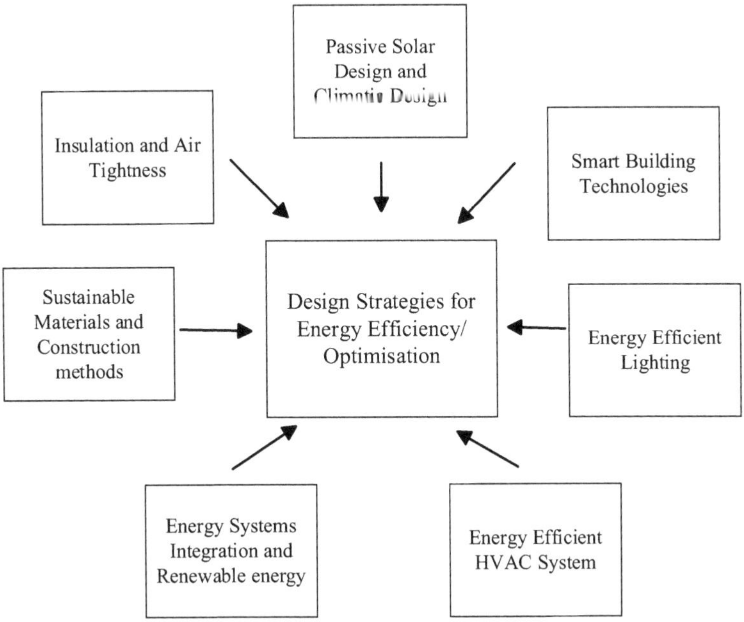

Fig. 1 Key design strategies for energy optimisation in buildings. *Source* Authors

A well-insulated building uses less energy for heating and cooling, lowering operational energy demands and associated carbon emissions [32, 85]. Moreover, techniques such as passive solar heating and natural ventilation can significantly reduce energy consumption [38], which will be further elaborated in Sect. 2.2. Hence, energy efficiency must be considered from the design phase through to building operation, ensuring adoption of comprehensive energy-efficient design strategies.

2.2 *Passive Design Strategies for Energy Efficiency*

Passive design strategies are critical for enhancing energy efficiency. These strategies align with global commitment to reducing carbon emissions and addressing climate change through sustainable building practices. Contemporary studies indicate that passive solar design can reduce energy consumption by 30% in residential buildings [3, 14]. By incorporating passive design strategies such as correctly orientating the building for optimum solar exposure, optimising insulation, reducing air leaks and selecting materials with suitable thermal mass, designers can significantly reduce reliance on energy-intensive mechanical systems and reduce operational carbon emissions [10, 78].

Passive design strategies mainly address thermal and visual comfort needs of occupants [77]. To fulfil thermal comfort needs, mechanical heating, ventilation and

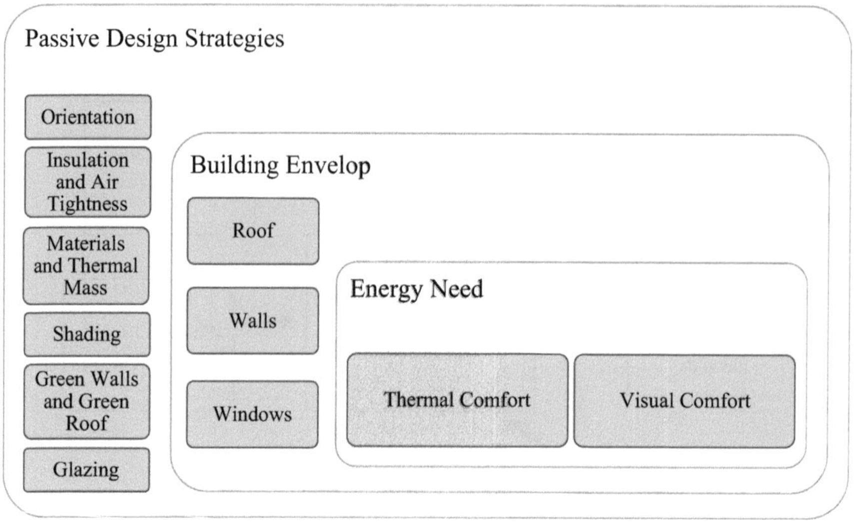

Fig. 2 Relationship between energy need, building envelop and passive design strategies. *Source* Authors

air-conditioning (HVAC) systems are widely used. In Australia, space heating and cooling accounts for around 40% of household energy use and around one-third in New Zealand [35]. Design of the building envelop, which includes the walls, roofs and windows plays a pivotal role in regulating temperature and minimising energy demand for heating and cooling systems [79]. Figure 2 illustrates how passive design strategies, such as proper orientation, insulation, material selection, thermal mass, shading, green walls/roofs and glazing, influence the building envelope (including the roof, walls and windows) to reduce energy needs, thereby achieving thermal and visual comfort.

For example, in Australia's temperate and subtropical climates, north-facing buildings maximise solar gain in winter and minimise heat gain in summer through appropriate types of shading strategies such as eaves, louvres or deciduous trees. This approach can significantly reduce the need for heating and cooling systems, cutting energy consumption by up to 50% in some climates [33]. Such passive design strategies not only support the transition to a low-carbon economy but also offer practical solutions for reducing operational costs and improving building performance in diverse climates.

2.3 *Active Design and Energy Systems Integration*

In addition to promoting passive design, the strategic use of active technologies and diverse energy systems in both new and existing dwellings can significantly improve a building's energy efficiency, thereby enhancing its overall sustainability [109]. Active technologies are the opposite of passive technologies, relying on mechanical systems and advanced technologies to control indoor environmental factors such as temperature, humidity and air quality. Heating, ventilation and air-conditioning (HVAC) system is a key element in the realm of active technologies. The demand for HVAC systems has been increasing in recent years due to global climate change and the impact of pandemics. People are increasingly relying on HVAC systems to provide a safe and comfortable indoor environment [9]. However, the increasing demand for HVAC systems has consumed a large amount of energy and, as a result, exacerbated climate change and the energy crisis, creating a vicious cycle [34]. Therefore, improving the energy efficiency of HVAC systems is crucial.

Energy-efficient HVAC systems are designed to reduce energy consumption without sacrificing indoor comfort [105]. Smart technology application can help to save energies in a further step for the energy-efficient HVAC system [13]. Smart technologies involve advanced technologies such as artificial intelligence (AI), the Internet of Things (IoT), big data and machine learning [57]. The application of these technologies helps in obtaining and analysing data [87]. Additionally, smart technologies provide optimised solutions for systems based on the data analysis result, enabling efficient resource utilisation, improving work efficiency and enhancing the overall user experience [26, 56].

Renewable energy integration also plays a significant role in carbon reduction. The use of renewable energy during the building operation period emits little to no carbon dioxide or other greenhouse gases, thus greatly reducing environmental pressure [58]. There are many available renewable energy technologies, such as solar power generation, solar thermal technology, wind turbines and geothermal systems. The integration of renewable energy technologies into building design can further enhance opportunities for utilising renewable energy [75]. For instance, solar panels can be integrated into various parts of the building envelope, including roofs and facades [96]. These integrated designs not only improve the efficiency of renewable energy use but also introduce a new aesthetic element different from traditional facade materials. In low-carbon building design, these technologies and integrated design methods should be considered to enhance building sustainability.

2.4 *Role of Smart Technologies for Energy Optimisation*

Smart technologies, including smart sensors and automation systems, play an important role in controlling energy usage during building operation. Smart sensors are useful for real-time monitoring and data collection of a building's physical environment, including temperature, humidity, lighting and occupancy [68]. These data are gathered by the sensors and then transferred to the automation system. After analysing the data, the automation system can adjust the building control system to optimise building performance. The adjustments made by the sensors and automation control system align energy demand with supply. In this way, energy efficiency can be improved without compromising user experience [40]. For example, in some office buildings, occupancy sensors can detect how many people are in a specific area. If the sensor detects that there are no people in the area, the automation control system will turn off the lights to save energy.

Building Information Modelling (BIM) is another technology to help optimising building energy performance. BIM is an integrated technology platform that visualises building information and facilitates seamless data exchange. The integrated technology platform includes software for design, building modelling, simulation, analysis and collaboration, spanning from the building design phase to the operation phase [49]. Using BIM technology, the performance of the building model can be simulated from multiple perspectives, including energy performance, physical environment, structural integrity, cost and construction sequence. Potential issues that may arise during construction or operation can be anticipated and resolved during the design phase [103]. This early optimisation of the building design helps to address problems before they occur, significantly improving resource efficiency.

Energy simulation is one of the BIM technologies that help reduce carbon emissions during building operation [41]. Energy simulation software can use building design information, occupancy patterns and climate characteristics to predict energy consumption during the building's operational phase [27]. These simulation results reflect various aspects of the building's energy usage, providing engineers with a basis for optimising the design. Typically, engineers make multiple adjustments to building parameters to understand how each factor impacts energy consumption. This process helps them identify the optimal combination of parameters to achieve the lowest possible energy use.

2.5 *Role of Building Materials in Enhancing Energy Efficiency*

The choice of building materials significantly influences the energy efficiency of buildings, impacting both their operational energy consumption and environmental footprint. Various studies highlight the importance of innovative and sustainable materials in achieving energy-efficient designs [1, 97]. While innovative materials offer promising solutions for energy efficiency, the complexity of building systems necessitates a holistic approach to material selection and design to maximise benefits [51].

As depicted in Fig. 3, building materials that contribute to energy efficiency are characterised by thermal insulation, thermal mass, reflectivity (Albedo), air tightness, durability and breathability, all of which help in optimising temperature regulation, reducing energy consumption and improving long-term sustainability [54, 106].

Integrating low-carbon materials, such as natural fibres, cross-laminated timber, recycled glass and plastic insulation and low-carbon concrete can substantially reduce the environmental impact of construction, while improving the thermal performance

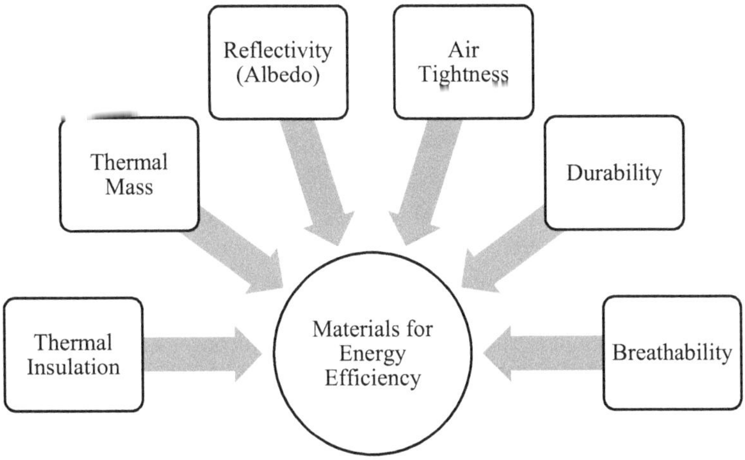

Fig. 3 Characteristics of building materials contributing to energy efficiency. *Source* Authors

of buildings [66]. Natural fibres such as straw, hemp, wool and flax are increasingly being used in construction for their excellent insulating properties and environmental benefits. These materials are compostable, less polluting than conventional insulation and help regulate indoor temperatures [46]. Such natural materials help in reducing heat transfer through walls, roofs and floors, thereby lowering the energy demand for heating and cooling systems. Additionally, using materials with a high thermal mass, such as low-carbon concrete and hemp brick, can regulate indoor temperatures by absorbing and releasing heat slowly, further enhancing energy efficiency.

Case studies have shown the significant impact of material selection on reducing operational carbon. For example, in Australia, projects utilising cross-laminated timber (CLT) have demonstrated reductions in both embodied and operational carbon compared to traditional steel and concrete construction [50]. The use of innovative materials like CLT not only lowers emissions during the construction phase but also provides superior insulation properties, reducing the building's energy needs over its lifetime. Hence, building materials not only play an important role in reducing operational carbon but also embodied carbon, which will be further discussed in Sect. 3.

3 Reducing Embodied Carbon Through Sustainable Design and Materials

3.1 Understanding Embodied Carbon in Building Design

Embodied carbon refers to the total greenhouse gas emissions associated with the production, transportation, installation, maintenance and disposal of building materials [19]. Unlike operational carbon, which is the energy consumed during a building's use, embodied carbon accounts for the emissions that occur before a building is even occupied. These emissions are significant, as they can contribute up to 50–75% of a building's total life cycle emissions [83]. Material selection, including the use of high-carbon materials like concrete and steel, plays a critical role in determining the embodied carbon footprint, along with construction methods that may either exacerbate or mitigate emissions. This makes reducing embodied carbon a key element in sustainable design strategies.

Addressing embodied carbon during the design phase is crucial for lowering a building's overall carbon footprint. Early-stage decisions regarding materials, structural systems and construction techniques have a lasting impact on the building's environmental performance [69]. For example, modular construction and prefabrication are methods that can minimise waste and reduce carbon emissions compared to traditional on-site construction processes [59]. Considering embodied carbon early allows architects and engineers to design buildings that align with net-zero goals and reduce life cycle emissions.

Incorporating embodied carbon considerations into the design process not only aligns with global sustainability goals but also ensures compliance with evolving building regulations and standards [90]. Many countries are beginning to include embodied carbon assessments in their building codes, as part of broader efforts to achieve carbon neutrality in the construction industry. Furthermore, accounting for embodied carbon can enhance a building's overall value by reducing future regulatory risks and improving long-term sustainability [84]. Therefore, a comprehensive understanding of embodied carbon is essential for building professionals committed to creating environmentally responsible and energy-efficient designs.

3.2 Factors to Consider When Selecting Construction Methods and Materials for Sustainability and Carbon Reduction

Building materials play a key role in improving energy efficiency and reducing carbon emissions [36]. When selecting construction methods and materials for sustainability and carbon reduction, several factors as shown in Fig. 4 must be considered: Building

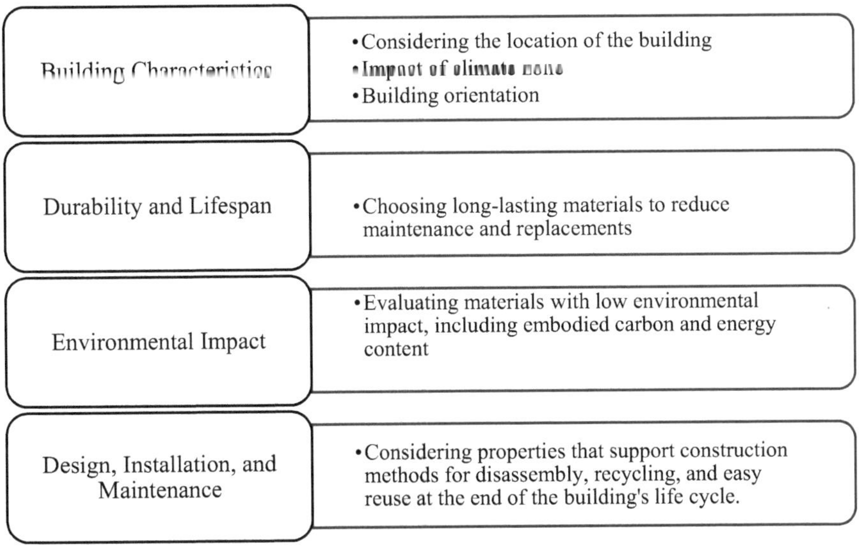

Fig. 4 Criteria for selecting materials and construction methods for sustainability and carbon reduction. *Source* Authors

Characteristics, Durability and Lifespan, Environmental Impact, and Design, Installation, and Maintenance [7, 20, 104]. Considering these factors ensures that buildings are both sustainable and energy-efficient while minimising carbon emissions throughout their life cycle.

Starting with the building's characteristics, the location of the building, the climate zone and building orientation directly affect a building's energy performance. In cooler climates, materials with high thermal mass such as brick or concrete can absorb and release heat slowly, helping regulate indoor temperatures without excessive heating or cooling. On the other hand, in warmer climates, building orientation and shading elements can be optimised to minimise solar gain, reducing the need for air-conditioning [41, 55]. Therefore, selecting construction materials that are climate-appropriate and enhance the building's thermal efficiency contributes significantly to lowering operational carbon over its lifespan.

Material durability is another critical factor in sustainable construction. Selecting long-lasting materials such as steel, stone or treated timber not only ensures the structure remains functional for a longer period but also minimises the frequency of repairs and replacements [36]. Reducing the need for frequent maintenance helps lower the cumulative embodied carbon associated with repair work and material production, thereby making a substantial contribution to sustainability.

Another factor to carefully evaluate when selecting building materials is their environmental impact of construction materials [107]. This evaluation should consider factors such as embodied carbon and energy content, as they significantly affect a building's overall sustainability. Opting for materials with low embodied energy, such as those that are recycled or involve less energy-intensive manufacturing processes, can considerably reduce the environmental footprint [89]. An example of this approach is the use of Environmental Product Declarations (EPDs), which provide transparent data on the environmental impact of construction materials. EPDs detail factors like embodied carbon, energy use and other ecological impacts throughout a product's life cycle, enabling architects and builders to make informed decisions [6]. For instance, choosing locally produced concrete with an EPD showing lower embodied carbon due to reduced transportation emissions and innovative manufacturing methods can significantly cut the total carbon emissions of a construction project [94]. By prioritising low-impact materials, construction projects can help lower the total carbon emissions linked to material sourcing and processing.

Finally, the factors of design, installation and maintenance play a crucial role in selecting sustainable building materials for carbon reduction. When choosing materials, it is important to consider their properties in relation to construction methods that support design for disassembly, dismantling and recycling, allowing for easier reuse or repurposing at the end of the building's life cycle [15]. The availability of technical skills for proper installation and maintenance also influences the long-term performance and sustainability of the materials [36]. Prioritising materials with renewable or reusable properties can further reduce resource consumption and waste. Additionally, selected materials should not only minimising emissions but also promote a healthy indoor environment for occupants, ensuring safety, durability and aesthetic appeal [2]. The robustness of materials, combined with a thoughtful design approach, can enhance the building's overall resilience and sustainability. This holistic approach leads to buildings that perform efficiently throughout their life cycle, aligning with broader carbon reduction goals.

3.3 Innovative Low-Carbon Materials and Technologies

The carbon emission coefficients of different construction materials vary, with steel and aluminium exhibiting significantly higher carbon emission coefficients compared to concrete and wood [39]. These materials with lower carbon emission coefficients are commonly referred to as low-carbon materials. Low-carbon materials are a new type of material that generates less carbon emissions throughout their lifecycle. Therefore, the use of low-carbon materials in the construction industry can help reduce its overall carbon emissions, representing a critical pathway towards achieving carbon neutrality [24]. With the evolving demands of the times and advancements in technology, the application of low-carbon materials is expected to contribute to the reduction of greenhouse gas emissions in the future. [53].

Bamboo and wood are excellent low-carbon materials, as they can absorb significant amounts of carbon dioxide during their growth, a process that facilitates carbon sequestration [25, 103]. From a life cycle perspective, the carbon dioxide absorbed by bamboo and wood exceeds the emissions generated during their production and processing [25, 102]. Another advantage of bamboo and wood is that they are renewable materials with faster regeneration rates compared to some non-renewable resources, which helps mitigate environmental degradation [48]. Due to their multiple advantages, bamboo and wood are used in various forms in building projects. For example, they can serve as structural materials, wall panels and finishes [8]. Because of their effective thermal and sound insulation properties, using bamboo and wood as wall panels and finishes enhances thermal and acoustic insulation, contributing to improved indoor comfort levels in buildings [95].

Traditional building materials can also be transformed into low-carbon alternatives through technological advancements [5]. One example is the development of green concrete, which can replace traditional concrete to help reduce carbon emissions. Green concrete reduces the amount of cement required, as cement production typically releases a significant amount of CO_2 [99]. Instead, green concrete incorporates alternative materials like fly ash and slag to replace cement [74]. Due to changes in the composition of green concrete, its carbon dioxide absorption rate is increased. Traditional concrete can only absorb a minimal amount of CO_2, which is insignificant compared to the CO_2 released during its production [5]. However, green concrete can absorb more CO_2 compared to traditional concrete due to the use of mineral admixtures, which enhance its carbon absorption capacity. Additionally, some types of green concrete can incorporate carbon injection and carbon sequestration technologies, further increasing their CO_2 absorption potential [108].

Building-integrated greening is also a form of carbon capture technology, as plants absorb carbon dioxide from the air through photosynthesis, effectively reducing CO_2 concentration in the atmosphere [21]. Greening systems often adopt modular design, consisting of planting substrates, modular structures, smart irrigation systems and waterproof layers [72]. When applied to the building facade, these greening modules are typically integrated with facade components to maximise the green area [96]. Building-greening integrated design is a low-carbon design method that offers multiple benefits. For instance, green roofs and green walls not only serve as carbon storage containers but also provide thermal insulation for buildings and help mitigate the urban heat island effect [100]. This implies that the application of greening technologies can reduce the demand for heating and cooling, thereby decreasing energy use and CO_2 emissions associated with building operation. Furthermore, greening systems also contribute to improving air quality and reducing noise pollution [71].

3.4 Assessments to Support Low-Carbon Design Decisions

Assessment tools play a crucial role in low-carbon design by providing scientific and comprehensive data, enabling engineers to make informed decisions that achieve low-carbon goals across various construction phases [62]. Various assessment tools can be employed in low-carbon design, including Life Cycle Assessment (LCA), Environmental Product Declarations (EPDs), and Whole-Building Carbon Assessments. This section will introduce these assessment tools.

LCA is a systematic method for low-carbon design that evaluates the environmental impacts of a project from raw material production, processing, transportation and usage through to disposal and recycling [70]. LCA could incorporate either a 'cradle-to-grave' or 'cradle-to-cradle' approach [17]. 'Cradle to grave' refers to a product's life cycle in which it ends up as waste after reaching the end of its useful life. In contrast, when a product is recycled or upcycled at the end of its life, the LCA model is referred to as 'cradle to cradle' [62]. The primary objective of using LCA is to understand the overall environmental impact of all aspects of the product. This evaluation method helps decision-makers gain a comprehensive view of a project's impact on the environment over its entire life cycle, rather than focusing on a single phase [76].

Generally, LCA consists of four stages: goal and scope definition, inventory analysis, impact assessment and interpretation [52]. In the first step, the purpose and boundaries of the assessment are defined, which is crucial as it forms the basis for subsequent steps, including data collection and model building. Following this, the inventory analysis provides quantitative information on the entire lifecycle of the product system, covering raw materials, energy, transportation, pollutants and waste. In the impact assessment phase, environmental models are developed based on data from the inventory analysis to quantify the product's environmental impacts. Finally, in the interpretation phase, results are analysed and interpreted and potential improvement measures are identified, which are essential for making environmentally beneficial decisions [45].

Compared to LCA, EPDs provide information on the environmental impacts of specific stages within a product's life cycle [18]. EPD information is typically provided by manufacturers, and the most common coverage of EPDs includes environmental impact indicators from raw material extraction to the production stage, excluding subsequent phases such as usage, maintenance and recycling [80]. Although EPDs do not offer as comprehensive information as LCA, they provide publicly accessible and reliable data for product comparison.

Whole-Building Carbon Assessments are tools used to evaluate the carbon emissions of a building throughout its life cycle. The primary objective is to analyse the impact of design decisions on life cycle carbon emissions, thereby assisting in optimising designs during the early stages of a building project to reduce life cycle carbon emissions [108]. Unlike LCA, Whole-Building Carbon Assessments focus primarily on embodied carbon and operational carbon, while LCA includes other environmental indicators beyond carbon emissions, such as acidification potential and eutrophication potential [43]. Consequently, Whole-Building Carbon Assessments reduce complexity and enhance practicality, making them well-suited for low-carbon building certifications.

There are also other digital tools for carbon modelling and assessment, including One Click LCA, Tally and EC3 (Embodied Carbon in Construction Calculator). These tools can simulate the impact of material selection on embodied carbon during the design phase, supporting the selection of suitable low-carbon materials [86]. Additionally, some tools, such as One Click LCA and Tally, can be used within BIM, allowing for immediate comparison of carbon emissions data from different material choices, which further enhances collaborative efficiency [65]. However, the extent to which these tools can be widely used in a country or region depends on the availability of data support, which may affect the adaptability of different tools across various countries [86].

4 Conclusion

This chapter emphasises the critical importance of integrating both active and passive design strategies to effectively reduce carbon emissions in construction. This implies that both active and passive design can contribute to carbon reduction, and focusing on only one aspect may limit the effectiveness of carbon reduction in construction projects. To reduce carbon emissions, digital technologies such as BIM can be applied to enhance building performance. Another emphasis fact of this chapter is that the application of innovative low-carbon materials demonstrates significant potential in decreasing embodied carbon emissions. Therefore, it is crucial to select and utilise suitable low-carbon materials for achieving low-carbon goals and to employ appropriate assessment tools, such as LCA, to inform decision-making.

The findings from this chapter align with existing literature on construction sustainability and extend the current body of knowledge. Compared to previous studies that focus solely on either passive or active design, this chapter provides insights into the relationship between low-carbon building design and the integration of both passive and active design. This study proposes a holistic approach to low-carbon design, which will help projects further reduce carbon emissions. Additionally, methods to reduce carbon emissions are represented in this study from two aspects, operational carbon and embodied carbon. Unlike previous research, which has predominantly focused on operational carbon, this chapter incorporates strategies for reducing the carbon footprint through the application of low-carbon materials and carbon emission assessment tools. These strategies are particularly effective for reducing carbon emissions over the entire life cycle of a building.

This chapter outlines whole life cycle strategies for low-carbon design. However, it does not explore how to prioritise and optimise these design elements when choices are limited. Given that most design projects cannot address all aspects equally, it is necessary to make choices under various constraints. Furthermore, suitable low-carbon design methods may vary across projects in different countries and geographic locations. Specific strategies need to be considered based on local conditions. Therefore, future research could focus on the following two areas to provide further guidance: First, analysing the sensitivity of various low-carbon design elements to carbon reduction. Second, develop more targeted low-carbon design methods adapted to the characteristics of individual countries and regions. For example, creating low-carbon material libraries tailored to different national contexts.

In the context of global climate change, this chapter, based on a critical literature review, presents carbon reduction design strategies for the construction industry across both operational and embodied carbon stages. These strategies encompass aspects such as design elements, energy optimisation, material selection, technology application and assessment tools. The findings indicate that low-carbon design should involve an integrated approach of both active and passive design strategies. The integration of active and passive design is essential for reducing carbon emissions in construction. Another key finding is that the application of innovative technologies and materials is critical to reducing the carbon footprint.

This chapter offers construction stakeholders such as building designers, engineers and policymakers clear insights into how to reduce carbon emissions throughout the entire life cycle, which will further promote the development of low-carbon construction. Additionally, this study extends the existing body of knowledge, potentially contributing to theoretical advancement, methodological innovation and practical applications. In the future, the application of innovative technologies and materials in sustainable projects will continue to be a key focus in the field of low-carbon research. It is anticipated that future theoretical and practical research will collectively drive the construction industry towards achieving carbon neutrality.

References

1. Abobakirov A (2023) Energy efficient building materials in the design of buildings and structures. Holders Reason 1(3):406–412
2. Alassaf Y (2024) Comprehensive review of the advancements, benefits, challenges, and design integration of energy-efficient materials for sustainable buildings. Buildings 14(9):2994
3. Alawode A, Rajagopalan P (2022) Feasibility of net zero energy high rise apartment buildings in Australia. Sol Energy 231:158–174
4. Al-Mudhaffer AF, Saleh SK, Kadhum GI (2022) The role of sustainable materials in reducing building temperature. Mater Today Proc 61:690–694
5. Althoe YF, Ansari WS, Sufian M, Deifalla AF (2023) Advancements in low-carbon concrete as a construction material for the sustainable built environment. In: Developments in the built environment, p 100284
6. Anderson J (2023) Reducing embodied carbon in the built environment: the role of environmental product declarations. Doctoral dissertation, The Open University, UK
7. Arroyo P, Tommelein ID, Ballard G (2016) Selecting globally sustainable materials: a case study using choosing by advantages. J Constr Eng Manag 142(2):05015015
8. Asif M (2009) Sustainability of timber, wood and bamboo in construction. Sustainability of construction materials. Elsevier
9. Asim N, Badiei M, Mohammad M, Razali H, Rajabi A, Chin Haw L, Jameelah Ghazali M (2022) Sustainability of heating, ventilation and air-conditioning (HVAC) systems in buildings—an overview. Int J Environ Res Public Health 19:1016
10. Athienitis AK, Santamouris M (2013) Thermal analysis and design of passive solar buildings. Routledge
11. Australian Building Codes Board (2022) Building for 7 stars: top tips and guidance. [Online] Available at: https://ncc.abcb.gov.au/news/2022/building-7-stars-top-tips-and-guidance. Accessed 6 Oct 2024
12. Australian Government (2024) Buildings—Department of Climate Change, Energy, the Environment and Water [Online] Available at: https://www.dcceew.gov.au/energy/energy-efficiency/buildings. Accessed 1 Oct 2024
13. Barone G, Buonomano A, Forzano C, Giuzio GF, Palombo A, Russo G (2023) A new thermal comfort model based on physiological parameters for the smart design and control of energy-efficient HVAC systems. Renew Sustain Energy Rev 173:113015

14. Bekele MT, Atakara C (2023) Residential building energy conservation in mediterranean climate zone by integrating passive solar and energy efficiency design strategies. Buildings 13(4):1073
15. Bertino G et al (2021) Fundamentals of building deconstruction as a circular economy strategy for the reuse of construction materials. Appl Sci 11(3):939
16. Brandão M et al (2013) Key issues and options in accounting for carbon sequestration and temporary storage in life cycle assessment and carbon footprinting. Int J Life Cycle Assess 18:230–240
17. Bui TT, Wilkinson S, Macgregor C, Domingo N (2023) Decision making in reducing carbon emissions for building refurbishment: case studies of university buildings in New Zealand. Build Environ 242:110557
18. Bui TTP, Macgregor C, Wilkinson S, Domingo N (2023) Towards zero carbon buildings: issues and challenges in the New Zealand construction sector. Int J Constr Manag 23:2709–2716
19. Cabeza LF, Boquera L, Chafer M, Verez D (2021) Embodied energy and embodied carbon of structural building materials: worldwide progress and barriers through literature map analysis. Energy Build 231:110612
20. Chan M, Masrom MA, Yasin SS (2022) Selection of low-carbon building materials in construction projects: construction professionals' perspectives. Buildings 12(4):486
21. Chhabra J, Rakha T (2020) Building-integrated carbon sequestration techniques: towards mitigating climate change. AIA/ACSA Intersections Research Conference: Carbon
22. Cordier S, Blanchet P, Robichaud F, Amor B (2022) Dynamic LCA of the increased use of wood in buildings and its consequences: integration of CO_2 sequestration and material substitutions. Build Environ 226:109695
23. Cusenza MA, Guarino F, Longo S, Cellura M (2022) An integrated energy simulation and life cycle assessment to measure the operational and embodied energy of a Mediterranean net zero energy building. Energy Build 254:111558
24. Das O, Restás Á, Shanmugam V, Sas G, Försth M, Xu Q, Jiang L, Hedenqvist MS, Ramakrishna S (2021) Demystifying low-carbon materials. Mater Circular Econ 3:1–11
25. De Araujo V, Vasconcelos J, Cortez-Barbosa J, Morales E, Christoforo A, Gava M, Lahr F, Garcia J (2020) Wood consumption and fixations of carbon dioxide and carbon from timber housing techniques: a Brazilian panorama. Energy Build 216:109960
26. De Assis Dornelles J, Ayala NF, Frank AG (2022) Smart working in industry 4.0: how digital technologies enhance manufacturing workers' activities. Comput Ind Eng 163:107804
27. Del Ama Gonzalo F, Moreno Santamaría B, Montero Burgos MJ (2023) Assessment of building energy simulation tools to predict heating and cooling energy consumption at early design stages. Sustainability 15:1920
28. De Wolf C, Dodd M, Byers B, Donatello S (2023) Whole life cycle environmental impact assessment of buildings: developing software tool and database support for the EU framework Level (s). In: Resources, conservation and recycling, vol 188
29. De Wolf C, Pomponi F, Moncaster A (2017) Measuring embodied carbon dioxide equivalent of buildings: a review and critique of current industry practice. Energy Build 140:68–80
30. Deo Prasad MD et al (2019) Race to net zero carbon: a climate emergency guide. Science
31. Echenagucia TM, Moroseos T, Meek C (2023) On the tradeoffs between embodied and operational carbon in building envelope design: the impact of local climates and energy grids. Energy Build 278:112589
32. Eicker U (2009) Low energy cooling for sustainable buildings. John Wiley & Sons
33. Elaouzy Y, Fadar A (2022) Energy, economic and environmental benefits of integrating passive design strategies into buildings: a review. Renew Sustain Energy Rev 167:112828
34. Elnagar E, Arteconi A, Heiselberg P, Lemort V (2024) Integration of resilient cooling technologies in building stock: impact on thermal comfort, final energy consumption, and GHG emissions. Build Environ 261:111666
35. Energy Rating (2024) Heating and cooling. [Online] Available at: https://www.energyrating.gov.au/consumer-information/products/heating-and-cooling#:~:text=Space%20heating%20and%20cooling%20appliances,big%20impact%20on%20energy%20costs. Accessed October 2024

36. Eze EC, Ugulu RA, Onyeagam OP, Adegboyega AA (2021) Determinants of sustainable building materials (SBM) selection on construction projects. Int J Constr Supply Chain Manage 11(2):166–194
37. Fang D, Brown N, De Wolf C, Mueller C (2023) Reducing embodied carbon in structural systems: a review of early-stage design strategies. J Build Eng 76:107054
38. Figueroa-Lopez A, Arias A, Oregi X, Rodriguez I (2021) Evaluation of passive strategies, natural ventilation and shading systems, to reduce overheating risk in a passive house tower in the north of Spain during the warm season. J Build Eng 43:102607
39. Ghanbari M (2023) Environmental impact assessment of building materials using life cycle assessment. J Architectural Environ Structural Eng Res 6:11–22
40. Gupta A, Badr Y, Negahban A, Qiu RG (2021) Energy-efficient heating control for smart buildings with deep reinforcement learning. J Build Eng 34:101739
41. Habibi S (2024) The effect of building orientation on energy efficiency. Clean Technol Environ Policy 26(4):1315–1330
42. Hafez FS et al (2023) Energy efficiency in sustainable buildings: a systematic review with taxonomy, challenges, motivations, methodological aspects, recommendations, and pathways for future research. Energ Strat Rev 45:101013
43. Hellmeister M (2022) Comparative life cycle assessment of embodied carbon and operational energy of different building systems. The University of Maine
44. Hestnes AG (1999) Building integration of solar energy systems. Sol Energy 67(4–6):181–187
45. Hollberg A, Kiss B, Röck M, Soust-Verdaguer B, Wiberg AH, Lasvaux S, Galimshina A, Habert G (2021) Review of visualising LCA results in the design process of buildings. Build Environ 190:107530
46. Indwar R, Titiksh A (2024) Organic wastes as a sustainable alternative to synthetic thermal insulating materials. In: Environment, development and sustainability, pp 1–41
47. Ingrao C et al (2018) How can life cycle thinking support sustainability of buildings? Investigating life cycle assessment applications for energy efficiency and environmental performance. J Clean Prod 201:556–569
48. Isukuru EJ, Ogunkeyede AO, Adebayo AA, Uruejoma MF (2023) Potentials of bamboo and its ecological benefits in Nigeria. Adv Bamboo Sci 4:100032
49. Jasiński A (2021) Impact of BIM implementation on architectural practice. Architectural Eng Des Manage 17:447–457
50. Jayalath A et al (2020) Life cycle performance of cross laminated timber mid-rise residential buildings in Australia. Energy Build 223:110091
51. Judkoff R (2008) Increasing building energy efficiency through advances in materials. MRS Bull 33(4):449–454
52. Kamari A, Kotula BM, Schultz CPL (2022) A BIM-based LCA tool for sustainable building design during the early design stage. Smart Sustain Built Environ 11:217–244
53. Kharissova AB, Kharissova OV, Kharisov BI, Méndez YP (2024) Carbon negative footprint materials: a review. Nano-Structures Nano-Objects 37:101100
54. Khosla R et al (2022) Sustainable cooling in a warming world: technologies, cultures, and circularity. Annu Rev Environ Resour 47(1):449–478
55. Kontoleon KJ, Eumorfopoulou EA (2010) The effect of the orientation and proportion of a plant-covered wall layer on the thermal performance of a building zone. Build Environ 45(5):1287–1303
56. Kothamali PR, Mandaloju N, Dandyala SSM (2022) Optimizing resource management in smart cities with AI. Unique Endeavor Bus Soc Sci 1:174–191
57. Li J, Herdem MS, Nathwani J, Wen JZ (2023) Methods and applications for artificial intelligence, big data, internet of things, and blockchain in smart energy management. Energy AI 11:100208
58. Li S, Cui Y, Banaitienė N, Liu C, Luther MB (2021) Sensitivity analysis for carbon emissions of prefabricated residential buildings with window design elements. Energies 14:6436
59. Liu G et al (2022) Greenhouse gas emissions management in prefabrication and modular construction based on earned value management. J Constr Eng Manag 148(6):04022034

60. Liu Z, Zhang X, Sun Y, Zhou Y (2023) Advanced controls on energy reliability, flexibility, resilience, and occupant-centric control for smart and energy-efficient buildings—a state-of-the-art review. In: Energy and buildings, p 113436
61. Min J et al (2022) The effect of carbon dioxide emissions on the building energy efficiency. Fuel Elsivier 326:124842
62. Minunno R, O'Grady T, Morrison GM, Gruner RL (2021) Investigating the embodied energy and carbon of buildings: a systematic literature review and meta-analysis of life cycle assessments. Renew Sustain Energy Rev 143:110935
63. Mofidi F, Akbari H (2020) Intelligent buildings: an overview. Energy Build 223:110–192
64. Moncaster AM, Song JY (2012) A comparative review of existing data and methodologies for calculating embodied energy and carbon of buildings. Int J Sustain Build Technol Urban Dev 3(1):26–36
65. Morsi DMA, Ismaeel WS, El Hamed A, Othman AAE (2020) Applying LCA-BIM integration for a sustainable management process. In: ARCOM 2020—Assoc Res Constr Manag 36th Annu Conf 2020—Proc, pp 416–424
66. Nasr Y et al (2023) Comprehensive review of innovative materials for sustainable buildings' energy performance. Energies 16(21):7440
67. New Zealand Ministry of Business, I. a. E. (2024) Building code compliance. [Online] Available at: https://www.building.govt.nz/building-code-compliance. Accessed 6 Oct 2024
68. Newton S, Shirazi A, Christensen P (2023) Defining and demonstrating a smart technology configuration to improve energy performance and occupant comfort in existing buildings: a conceptual framework. Int J Build Pathol Adaptation 41:182–200
69. Nielsen AN, Jensen RL, Larsen TS, Nissen SB (2016) Early stage decision support for sustainable building renovation—a review. Build Environ 103:165–181
70. Nidhin BKSN, Domingo N, Bui TTP, Wilkinson S (2023) Construction stakeholders' knowledge on zero carbon initiatives in New Zealand. Int J Build Pathol Adaptation
71. Nieuwenhuijsen MJ (2021) New urban models for more sustainable, liveable and healthier cities post covid-19; reducing air pollution, noise and heat island effects and increasing green space and physical activity. Environ Int 157:106850
72. Ogut O, Tzortzi NJ, Bertolin C (2022) Vertical green structures to establish sustainable built environment: a systematic market review. Sustainability 14:12349
73. Omar O (2018) Intelligent building, definitions, factors and evaluation criteria of selection. Alex Eng J 57(4):2903–2910
74. Orozco C, Babel S, Tangtermsirikul S, Sugiyama T (2024) Comparison of environmental impacts of fly ash and slag as cement replacement materials for mass concrete and the impact of transportation. Sustain Mater Technol 39:e00796
75. Østergaard PA, Duic N, Noorollahi Y, Mikulcic H, Kalogirou S (2020) Sustainable development using renewable energy technology, vol 146, pp 2430–2437
76. Pisini S, Thammadi S, Wilkinson S (2023) Embodied carbon emissions for net-zero carbon buildings: a comprehensive study of New Zealand. In: International conference on engineering, project, and production management, pp 739–755
77. Rabani M, Madessa HB, Nord N (2021) Achieving zero-energy building performance with thermal and visual comfort enhancement through optimization of fenestration, envelope, shading device, and energy supply system. Sustain Energy Technol Assess 44:101020
78. Ralegaonkar RV, Gupta R (2010) Review of intelligent building construction: a passive solar architecture approach. Renew Sustain Energy Rev 14(8):2238–2242
79. Rana K (2021) Towards passive design strategies for improving thermal comfort performance in a naturally ventilated residence. J Sustain Architecture Civil Eng 29(2):150–174
80. Rangelov M, Dylla H, Mukherjee A, Sivaneswaran N (2021) Use of environmental product declarations (EPDs) of pavement materials in the United States of America (USA) to ensure environmental impact reductions. J Clean Prod 283:124619
81. Rathnam PM, Ram S (2022) Imbibing energy efficiency in buildings through sustainable materials—a review. In: Advances in sustainable materials and resilient infrastructure, pp 399–414

82. Rissman J et al (2020) Technologies and policies to decarbonize global industry: review and assessment of mitigation drivers through 2070. Appl Energy 266:114848
83. Robati M et al (2021) Carbon value engineering: a framework for integrating embodied carbon and cost reduction strategies in building design. Build Environ 192:107620
84. Roberts M, Allen S, Coley D (2020) Life cycle assessment in the building design process—a systematic literature review. Build Environ 185:107274
85. Sadineni SB, Madala S, Boehm RF (2011) Passive building energy savings: a review of building envelope components. Renew Sustain Energy Rev 15(8):3617–3631
86. Sajid ZW, Khan SA, Hussain F, Ullah F, Khushnood RA, Soliman N (2024) Assessing economic and environmental performance of infill materials through BIM: a life cycle approach. In: Smart and sustainable built environment
87. Sarker IH (2021) Data science and analytics: an overview from data-driven smart computing, decision-making and applications perspective. SN Comput Sci 2:377
88. Simonen K, Rodriquez BX, De Wolf C (2017) Benchmarking the embodied carbon of buildings. Technology|Architecture + Design 1(2). https://doi.org/10.1080/24751448.2017.1354623
89. Sizirici B et al (2021) A review of carbon footprint reduction in construction industry, from design to operation. Materials 14:206094
90. Skillington K, Crawford RH, Warren-Myers G, Davidson K (2022) A review of existing policy for reducing embodied energy and greenhouse gas emissions of buildings. Energy Policy 168:112920
91. Soares N et al (2017) A review on current advances in the energy and environmental performance of buildings towards a more sustainable built environment. Renew Sustain Energy Rev 77:845–860
92. Sutkowska M et al (2024) Recent advances in prefabrication techniques for biobased materials towards a low-carbon future: from modules to sustainability. J Build Eng 109558
93. Tripanagnostopoulos Y (2014) New designs of building integrated solar energy systems. Energy Proc 57:2186–2194
94. Waldman B, Huang M, Simonen K (2020) Embodied carbon in construction materials: a framework for quantifying data quality in EPDs. Build Cities 1(1)
95. Wang J, DeMartino C, Xiao Y, Li Y (2018) Thermal insulation performance of bamboo-and wood-based shear walls in light-frame buildings. Energy Build 168:167–179
96. Wang W, Yang H, Xiang C (2023) Green roofs and facades with integrated photovoltaic system for zero energy eco-friendly building—a review. Sustain Energy Technol Assess 60:103426
97. Wang X et al (2022) A critical review on phase change materials (PCM) for sustainable and energy efficient building: design, characteristic, performance and application. Energy Build 260:111923
98. Wang X, Ramakrishnan S (2021) Environmental sustainability in building design and construction. Springer International Publishing
99. Wen J, Wang B, Dai Z, Shi X, Jin Z, Wang H, Jiang X (2023) New insights into the green cement composites with low carbon footprint: The role of biochar as cement additive/alternative. Resources Conserv Recycl 197:107081
100. Wong NH, Tan CL, Kolokotsa DD, Takebayashi H (2021) Greenery as a mitigation and adaptation strategy to urban heat. Nat Rev Earth Environ 2:166–181
101. Wu Z et al (2023) Co-optimization of building energy systems with renewable generations combining active and passive energy-saving. Appl Energy 351:121514
102. Xu P, Zhu J, Li H, Wei Y, Xiong Z, Xu X (2022) Are bamboo construction materials environmentally friendly? A life cycle environmental impact analysis. Environ Impact Assess Rev 96:106853
103. Xu X, Xu P, Zhu J, Li H, Xiong Z (2022) Bamboo construction materials: Carbon storage and potential to reduce associated CO_2 emissions. Sci Total Environ 814:152697
104. Yahia AK, Rahman DM, Shahjala M, Morshed AS (2024) Sustainable materials selection in building design and construction. Int J Sci Eng 1(4):106–119

105. Yang Y, Hu G, Spanos CJ (2021) Stochastic optimal control of HVAC system for energy-efficient buildings. IEEE Trans Control Syst Technol 30:376–383
106. Zaera-Polo A, Anderson J (2021) The ecologies of the building envelope: a material history and theory of architectural surfaces. Actar D, Inc
107. Zari MP (2019) Ecosystem services impacts as part of building materials selection criteria. Mater Today Sustain 3:100010
108. Zhang S, Yuan Q, Ni J, Zheng K, Xu Y, Zhang J (2023) CO_2 utilization and sequestration in ready-mix concrete—a review. In: Science of the total environment, p 168025
109. Zong C, Chen X, Deghim F, Staudt J, Geyer P, Lang W (2024) A holistic two-stage decision-making methodology for passive and active building design strategies under uncertainty. Build Environ 251:111211

Supply Chain and Construction Practices for Low Carbon Construction

Namal Gamage, Srinath Perera, and Samudaya Nanayakkara

Abstract The construction sector plays a pivotal role in the economic development of nations, acting as a major contributor to global GDP and employment. However, the construction industry and its supply chain account for around 39% of global CO_2 emissions, yet it is challenging to minimise due to its complex and fragmented nature. Understanding the practices that assist low-carbon construction is imperative to reduce excessive emissions, yet limited attention has been received from academia and industry. Thus, this chapter examines carbon emissions in the construction supply chain and practices to be carried out for low-carbon construction. The initial part of the chapter discusses the construction supply chain, its characteristics, and low-carbon construction. The findings of the chapter are dedicated to carbon emissions and practices that can be adopted for low-carbon construction at different stages of the building life cycle. The findings reveal that embracing renewable energy sources, utilising energy-efficient machinery and equipment, and following policies, standards, and regulations are key low-carbon practices within the cradle-to-gate system boundary. Further, local sourcing, intermodal transportation, and optimising logistical operations were posed as significant practices for reducing carbon emissions associated with transport and logistics management. Moreover, off-site construction, managing waste through recycling and reusing, and design optimisation are vital practices to adopt at construction sites to drive towards low-carbon construction.

1 Introduction

Climate change has become a significant environmental challenge worldwide. Releasing greenhouse gases (GHGs) into the atmosphere is a major contributor to global warming, which in turn affects climate change [19]. The report, issued by the Intergovernmental Panel on Climate Change (IPCC), declared that GHG emissions should be reduced by 45% by 2030 compared to 2010 to achieve 100% reduction

---

N. Gamage · S. Perera (✉) · S. Nanayakkara
Centre for Smart Modern Construction (c4SMC), School of Engineering, Design and Built Environment, Western Sydney University, Sydney, Australia
e-mail: srinath.perera@westernsydney.edu.au

N. Domingo et al. (eds.), *Global Net Zero Carbon Practices in Construction*, Green Energy and Technology, https://doi.org/10.1007/978-981-96-3853-6_7

by 2050 [13]. Regarding GHG emissions, the construction industry is the worst offender due to its continuous heavy fossil fuel consumption [19]. Compared to other sectors, the construction industry plays a critical role in response to climate change, accounting for around 39% of energy-related emissions globally [13]. Therefore, minimising carbon emissions in the construction industry is one of the critical concerns in the present context [6].

The construction industry plays a pivotal role in the economic development of nations, acting as a significant contributor to global GDP and employment. It encompasses a wide range of activities including residential, commercial, and infrastructure projects. The rapid growth of the population, with around 83 million people added yearly, has led to an exceptional pace of urbanisation [37]. Thus, the construction industry is experiencing an unmatched phase of expansion worldwide [38], resulting in higher carbon-related emissions daily. The construction industry is well known for its diversified and complex supply chain, heavy fossil fuel consumption, and prevailing conventional practices. Hence, reducing carbon emissions in the construction industry has become challenging. Therefore, the construction supply chain (CSC) needs to be vigilantly overlooked in lowering carbon emissions [53]. Despite the significance, there is a limited knowledge on carbon emissions in the different phases of the CSC and practices to minimise them. Thus, this chapter examines the carbon emissions in the CSC along with the practices to reach towards low-carbon construction. Following the introduction, the literature review, research methodology, findings and discussion, and conclusion and further research are presented in subsequent sections.

2 Literature Review

2.1 Construction Supply Chain and Its Characteristics

A supply chain is a structure that unites all the processes involved in creating, distributing, and delivering products or services from their point of origin to final users [61]. It reflects the movement of data and resources through multiple phases, commencing with acquiring raw materials and ending with delivering the finished good or service [15]. The supply chain comprises several connected parts and starts with acquiring raw materials or components from suppliers [41]. Subsequently, these materials undergo transformations during manufacturing or production processes, creating products with enhanced value [39].

The construction industry involves complex and extensive project-based production such as buildings, bridges, and tunnels [7, 35]. Construction projects require the coordination of labour, machinery, components, materials, and specialists across multiple locations, resulting in dynamic and lengthy supply chains [45]. Thus, a CSC emerges as an intricate interconnection of individuals, companies, resources, activities, and technology involved in producing construction outputs such as buildings.

Each construction project has numerous supply chains that are dynamic and networked, and they include multiple tiers of manufacturers, suppliers, subcontractors, and sub-subcontractors [7, 14, 41]. Figure 1 illustrates the flow of information, money, raw materials, equipment, products, and services of the CSC.

The construction supply chain starts with mining raw materials, which are supplied to the manufacturers subsequently. They produce construction materials and equipment. Later, these resources are disseminated to sub-subcontractors, subcontractors, and the main contractor in the sequence of supply. Further, the raw materials can be directly supplied to the subcontractors, sub-subcontractors, and main contractor as required. Upon successful completion, the main contractor handovers the project to the client.

The information flow occurs bidirectionally, encompassing communication from the client to the manufacturer and vice versa (e.g. the requirement, material quantities, quality, etc.) [26]. The cash flow occurs from the client towards the manufacturer,

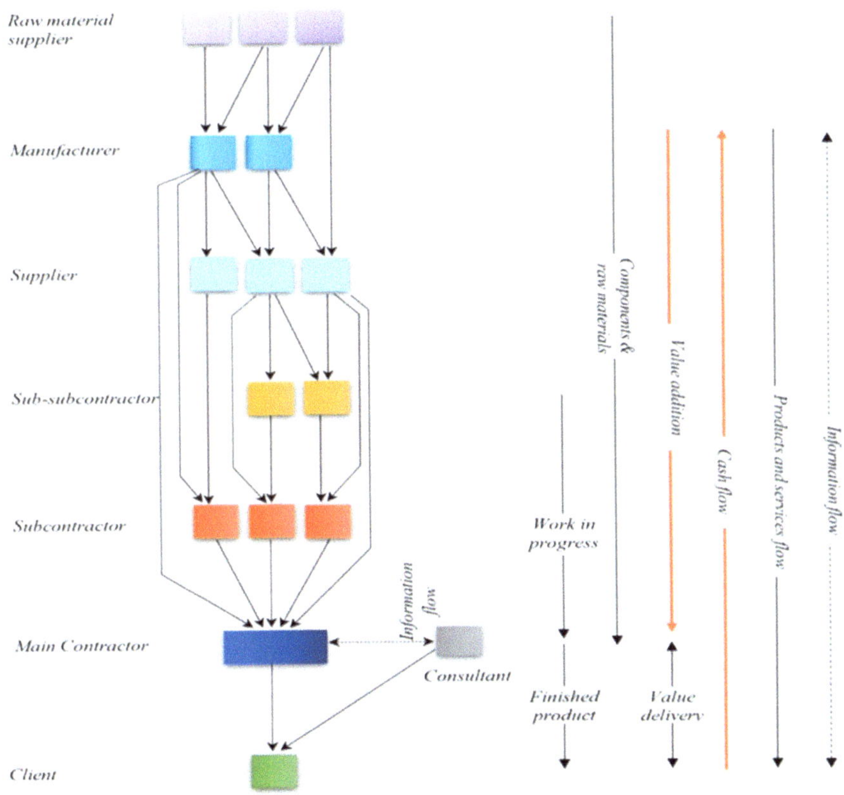

Fig. 1 Arrangement of the construction supply chain. *Source* [41]

while the flow of materials, equipment, and value addition occurs in the opposite direction [59]. The value chain constitutes a sequential process that enhances the value of a product at each stage, starting from the initial raw materials and culminating with the end product [43]. It involves extracting and manipulating raw materials through various stages and processes until a final product is created [39]. The value chain helps to identify the activities that generate significant value and allows for a comprehensive understanding of economic linkages within a construction project [36]. Following the discussion on the characteristics, carbon emissions in the CSC are discussed in subsequent sections.

2.2 Low-Carbon Construction

Carbon emissions within the building life cycle can be classified into two types: operational carbon and embodied carbon. Operational carbon refers to the CO_2 and equivalent gases that are emitted during the operational works of a building, like lighting, heating, and cooling [48]. Embodied carbon is the carbon-related emissions (CO_2 and equivalent gases) due to non-operational works of a building, such as material extraction, production, transportation, and construction, together with the downstream works like maintenance and demolition [7]. In contrast to operational carbon, embodied carbon emissions occur within each stage of the building life cycle. These stages are known as system boundaries and are presented in Fig. 2. These system boundaries are adopted in estimating carbon emissions within the construction industry.

Numerous efforts have been made to minimise operational carbon emissions; however, the investigation of embodied carbon remains limited due to the complex nature of the construction industry [5]. Attempting low-carbon projects is a recent

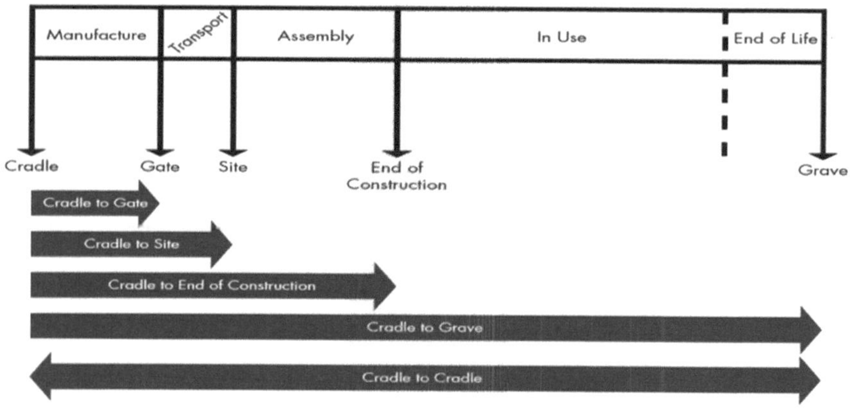

Fig. 2 System boundaries for estimating embodied carbon. *Source* [7]

trend in the construction sector [10], therefore, embodied carbon within the CSC becomes significant in reaching a low-carbon built environment.

Low-carbon construction refers to the development that carries out by minimising carbon-related emissions through appropriate measures [63]. In 2013, the low-carbon route map for the UK Built Environment, set out by the Green Construction Board, revealed that 39% of embodied carbon reduction needs to be achieved by 2050 compared to 2010 [24]. Utilising alternative materials with fewer emissions is one of the most critical steps in moving towards a low-carbon built environment. While the world is searching for solutions to minimise global warming, innovative designs and advanced construction methods with low emissions are becoming popular. Further, policymakers are promoting low-carbon construction practices to reach sustainable development goals [23]. Therefore, knowing the relevant practices to move towards low-carbon construction is imperative. The findings section presents carbon emissions within the different stages and appropriate practices to minimise them.

3 Research Methodology

The research methodology in this study involved a systematic literature search applying search rules to filter articles from the Scopus and Web of Science (WoS) databases. Subsequently, a screening was conducted to identify studies on carbon emissions in the construction industry. Thirty-one papers were selected from the screening, and this chapter was developed using them. A similar approach has been adopted by [49, 51] in their studies.

4 Findings and Discussion

4.1 Carbon Emissions During Cradle-to-Gate and Low-Carbon Practices

The cradle-to-gate system boundary accounts for significant emissions, which may exceed up to 80% of the overall emissions of a construction project. Raw material extraction and product manufacturing are key interrelated activities that must be scrutinised in conserving carbon emissions within the cradle-to-gate system boundary [57].

Raw materials such as sand, gravel, gypsum, stone, and lime are fundamental for the construction of buildings and infrastructure. Yet, little focus is given towards the sustainability of their extraction processes [66]. The extraction activities of traditional construction materials like cement, brick, and steel tend to have a higher negative impact on the environment [8]. Circularity Gap Report 2020 highlights that around 100 billion tonnes of raw materials are extracted each year, of which nearly 30%

represent gravel, sand, and crushed stone, predominantly utilised by the construction sector. Moreover, raw material extraction consumes massive amounts of fossil fuel and electricity, therefore, there are enormous CO_2 emissions worldwide [62].

Wu et al. [64], a study of carbon emissions in the construction industry of China, found that around 2.60 billion tonnes of CO_2 were released during the raw material extraction in 2015, having increased steadily over the past 15 years. While the environmental impacts during raw material extraction are becoming apparent, their contribution to climate change and the resulting GHG emissions have been lightly overlooked [62, 69]. The extracted raw materials are ultimately transformed into construction materials through a variety of production processes. According to [18], 39% of total CO_2 emissions are caused by the construction industry, and at least 11% are due to manufacturing materials and related products. It was also found that steel, aluminium, and cement are three key contributing materials to CO_2 emissions [64].

Steel manufacturing is a highly energy and carbon-intensive process, with a global average CO_2 emission of 1839 kg per tonne of steel produced. The steel sector consumes 5.9% of the global energy and is responsible for 6–9% of global CO_2 emissions. 70% of the steel is produced by the blast furnace using coal as a raw material for the process and as a source of energy. The use of coal is essential in steelmaking; therefore, significant carbon emissions are generated [44]. Following steel, using concrete in construction also predominantly affects carbon emissions worldwide.

The production of concrete and mortar has increased four times between 1990 and 2020 across the globe, resulting in an annual volume of approximately 26 gigatonnes [4]. According to [30], the main reason for the higher contribution of CO_2 in traditional concrete production is the substantial energy required to produce cement, which is the primary binding agent in concrete. Nassar et al. [42] pointed out that approximately 0.9 tonnes of CO_2 is emitted from the production of each tonne of cement. Moreover, data indicates that the cement industry is responsible for around 7% of CO_2 emissions worldwide [16].

Aluminium is another highly utilised material for constructing windows, facades, doors, staircases, and many other components in the building sector. It is also refined and produced through a highly energy-intensive process with a massive carbon footprint. As per the statistics, the aluminium industry itself accounts for around 1% of the global GHG emissions [22].

Several diverse practices can be adopted to minimise carbon emissions throughout the cradle-to-gate system boundary as elaborated below.

- Using renewable energy sources

Replacing fossil fuels with renewable energy, such as solar and wind, can significantly decrease the carbon footprint of both extraction and manufacturing processes. For example, mining operations can be powered by solar panels installed on site, reducing reliance on diesel generators. Additionally, using hydrogen and biomass as energy sources for material production, such as steelmaking, aids in the substantial reduction of CO_2 emissions [68]. Identifying the significance of moving towards renewable energy, China has developed a strategic plan, "Made in China 2025", to reach a 40% reduction of carbon emissions in the manufacturing industry by 2025 compared to 2015 [65].

- Utilising energy-efficient machinery and equipment

Using energy-efficient machinery in the manufacturing phase of construction significantly reduces carbon emissions. These advanced machines consume less energy to perform the same tasks, lowering the overall energy demand [32]. Moreover, upgrading and replacing older machinery and equipment contributes to substantial energy savings, ultimately minimising carbon emissions [1]. Additionally, proper maintenance increases the efficient life span of machinery, decreasing energy loss and reducing carbon emissions [56].

- Utilising recycled materials instead of virgin materials

While around 100 Bn tonnes of raw materials are extracted annually, 90% of them are virgin resources [62]. Thus, utilising recycled materials plays a significant role in reducing carbon emissions in the built environment. For example, using scrap metal (recycled metal) to produce steel members is a well-known method to reduce carbon emissions from steel production [27]. Gan et al. [21] indicated that using 30–100% of recycled scrap steel could reduce Embodied carbon emissions by 10–40% in reinforced concrete buildings. Additionally, recycled aluminium can save up to 95% of emissions compared to primary aluminium production, which uses virgin materials [22].

- Following regulations, policies, and standards

Following regulations, policies, and standards is pivotal to minimising carbon emissions. Illegal material extraction occurs frequently worldwide, and many adopt environmentally harmful practices, causing massive carbon emissions. Hence, strengthening the legal framework is essential to prevent these violations [62]. Next, an Environmental Product Declaration (EPD) outlines the environmental performance of construction products and services. EPDs detail various environmental parameters, including greenhouse gas emissions, from raw material extraction to releasing the final product from the factory gate and beyond [46]. Issuing EPD documents by the manufacturers facilitates the procurement process to identify the low-carbon products, which will eventually assist in minimising carbon emissions in the built environment.

- Utilising novel technologies

The construction industry is well known for its prevailing traditional practices throughout its life cycle. Most of these conventional processes consume massive amounts of fossil fuel and electricity, which causes abundant carbon emissions. Adopting novel technologies for material extraction and manufacturing processes minimises emissions [52]. Zhang et al. [68], a review conducted on the steel industry, pointed out that utilising top gas recycling (TGR) technology for steel manufacturing can effectively reduce carbon emissions by 26%. Moreover, carbon capture and storage (CCS) technology can capture carbon emissions during material production, thereby significantly reducing the CO_2 emissions released into the atmosphere. It is predicted that this technology has the capability to reduce 17% of global CO_2 emissions by 2050 [52].

4.2 Carbon Emissions During Gate-to-Site and Low-Carbon Practices

The gate-to-site system boundary involves transport and logistics management of construction materials, which play a significant role in the carbon footprint of the CSC [55]. The construction is highly reliant on transportation, with an average site receiving up to 8–10 tonnes of materials daily. However, relatively little attention has been paid to emissions associated with transporting goods to and from construction sites [25].

Sezer and Fredriksson [55] highlighted that optimising construction transportation and logistics management aids in improving both the productivity and sustainability of the industry. Further, the same study points out that construction projects with better environmental ratings have generated less emissions from transport and logistics. Predominant emissions within the gate-to-site system boundary are pertinent to transportation, however, the exact numbers are still debated. According to [11], materials transport constitutes 16% of the total emissions of a project, while [55] pointed out that around 10% of the emissions stem from transportation. However, in contrast, [54] mentioned that the transportation of materials to the site accounts for only 2.4% of the total CO_2 emissions of the project. The high deviations in transportation-related emissions may occur due to variations in transportation distances, modes, load efficiency, etc. [48]. Moreover, estimating transportation-related carbon emissions is challenging due to the difficulty of gathering data [55].

The rapid growth of the population has led to an increase in construction projects [33], which in turn has elevated transportation and logistics-related emissions. Thus, reducing these emissions is crucial to reaching sustainable development. Multiple practices can be adopted to minimise carbon emissions within the gate-to-site system boundary, as elaborated below.

- Local sourcing

Using local resources in construction projects significantly reduces carbon emissions related to the gate-to-site phase. By sourcing materials locally, transportation distances are minimised, leading to lower fuel consumption and fewer greenhouse gas emissions [27]. In proving it, [17] found that using local stones like sandstone, granite, and slate in the UK can significantly reduce Embodied carbon by 2% to 84% compared to importing stones from Spain, India, and China.

- Utilising alternative energy sources for vehicles

Karlsson et al. [29], a study conducted on Swedish road construction, pointed out that heavy-duty transportation-related GHGs accounted for 7% of the overall emissions of the project. Heavy-duty transportation frequently uses vehicles such as dump trucks, and flatbed trucks, which consume higher amounts of fossil fuels like diesel. Therefore, minimising associated emissions is challenging. It is a known fact that even though heavy-duty trucks constitute only 1% of all vehicles, they are responsible for 25% of total emissions from vehicles [9]. Utilising electric and hybrid vehicles can substantially reduce carbon emissions compared to diesel/petrol-powered vehicles. In supporting this, [29] highlighted in their study that utilising hybrid trucks for transportation can reduce more than 30% of the carbon emissions compared to diesel-powered trucks.

- Varying the mode of transportation

The impact of material transportation on carbon-related emissions varies significantly depending on the chosen mode. Different transportation methods can contribute to an increase in total carbon emissions, ranging from 0.04% to 8.4% [47, 67]. Three primary methods commonly used for transporting construction materials are truck, rail, and marine shipping [40]. Statistics show that marine transport generates only 26% of the emissions produced by truck transportation [12, 40]. Therefore, promoting intermodal transportation, which involves using multiple modes of transport (such as rail, road, and sea) within a single supply chain, can optimise the strengths of each mode and reduce overall emissions. For instance, materials can be transported by rail or ship for most of the journey and then transferred to trucks for more flexible, short-distance logistical operations.

- Optimising logistical operations

Logistical operations can be optimised through proper planning, coordination, and implementing advanced technologies to minimise carbon emissions. Prevailing transportation within the gate-to-site is frequently inefficient due to numerous reasons, such as low fill rates, empty running, and oversized vehicles leading to unnecessary emissions worldwide [34, 55]. Detailed planning, including efficient delivery schedules and route optimisation reduces fuel consumption, avoiding unnecessary trips [55, 58]. Therefore, associated emissions can be minimised.

- Implementing and following policies on managing transport and logistics

Implementing and following policies and standards can significantly reduce carbon emissions within the gate-to-site system boundary. Promoting local sourcing and energy-efficient transportation through policies minimises fossil fuel consumption, reducing carbon emissions. Moreover, mandating regular maintenance of vehicles and meeting emission standards can lower the GHGs emitted in transportation. While highlighting the importance of policies, construction logistics setup (CLS) is a governance structure established to oversee, manage, and monitor the flow of materials in a construction project [55]. Guerlain et al. [25] highlighted that practising according to the CLCs can significantly decrease CO_2 emissions from construction transportation by up to 30%.

4.3 On-Site Carbon Emissions and Low-Carbon Practices

A substantial portion of the overall carbon footprint of a construction project can be attributed to its on-site emissions that occurred during the construction phase. Fuel-powered machinery, equipment, vehicles, and high electricity consumption are well-known contributors to these on-site carbon emissions [6]. According to [54], a study on an apartment project in Korea with a construction area of 1934 m^2 containing a reinforced concrete structure found that carbon emissions from on-site construction accounted for 4.2% of the total upfront (cradle to end of construction) emissions. However, [50] researched a 180.4 m^2 wooden single-family house and found that the construction stage represents around 13% of upfront emissions. Therefore, on-site carbon emissions can differ significantly depending on variations in location, type of structure, and many other parameters.

Multiple practices can be adopted to minimise carbon emissions during on-site activities, leading to low-carbon construction. Some of them have been elaborated below.

- Utilising off-site construction

Off-site construction involves manufacturing components under quality-controlled factory conditions and assembling them during the construction phase at a site (Peiris et al. 2023). Prefabrication (off-site construction) has been gaining the attention of multiple industries, including construction, as a promising sustainable alternative with less carbon emissions. [60] systematically reviewed studies conducted on prefabricated buildings. The study concluded that the carbon emissions of prefabricated buildings range from 105 to 864 $kgCO_2/m^2$, which results in an average of 15.6% embodied carbon reduction compared to traditional buildings. Therefore, incorporating prefabrication into construction projects can be regarded as one of the most effective strategies for reducing carbon emissions.

- Using renewable energy sources for construction

The construction industry heavily depends on machinery, equipment, and construction methods that consume non-renewable energy sources, leading to massive emissions. The use of green energy sources, such as solar or wind energy, and fuel alternatives like hydrogen, can potentially reduce carbon emissions in the construction phase [53]. Therefore, the construction sector can benefit by investigating the possibility of replacing traditional on-site energy sources with renewable ones. For instance, construction sites can opt for sustainable solutions such as temporary photovoltaic systems, electric vehicles, and hydro-power harnessing [6].

- Upgrading and replacing old machinery

Construction sites utilise a wide variety of machinery and equipment, such as excavators and specialised vehicles such as cranes and forklifts. [31] recorded the on-site carbon emissions of 24 types of construction machinery and equipment and estimated the average emissions of these equipment to be 429.68 tCO_2e. However, the overall carbon emissions can increase substantially if construction machinery and equipment are not maintained regularly. Hence, periodic monitoring and upgrading of them is a vital practice to reduce excessive carbon emissions.

- Managing waste through recycling and reuse

An abundance of waste is generated from the construction industry and is responsible for around 35% of global landfill waste [28]. Recycling and reusing construction and demolition waste for construction can substantially reduce carbon emissions by minimising the usage of novel materials [2]. Depending on the type of construction and demolition waste, between 50 and 95% of it can be recycled [3]. For instance, steel recycling is regarded as one of the best end-of-life strategies for construction materials due to its significant benefits, including a reduction of up to 60% in embodied carbon and the ability to reuse nearly 100% of steel scrap into virgin steel [40]. Next, the concept of reuse is not limited to construction materials, building components can be designed in such a way that they can be disassembled and reused. [40] highlighted that building components such as panels and prefabricated structures can be reintegrated into the market after disassembly to reduce unwanted emissions due to new construction.

- Design optimisation

Design optimisation can be carried out throughout a construction project, from conceptual designing to ongoing construction. Hu [27] discussed several diverse design optimisation strategies and their potential to reduce carbon emissions. For example, using high-strength concrete minimises the size of structural elements, which in turn reduces carbon emissions. Increasing concrete strength twofold can decrease emissions by over 11% in reinforced concrete buildings [21]. Similarly, optimising the shape of the concrete beams can reduce carbon emissions by up to 38% compared to conventional designs [27]. Another strategy for design optimisation is to adaptively reuse existing structures instead of demolishing them and constructing new ones [20]. This approach will save substantial emissions due to the reduction of novel construction.

5 Conclusion and Further Research

The carbon emissions of the construction industry and its supply chains are known to be substantial and difficult to minimise due to their intricate nature. Thus, this chapter examines the characteristics of the construction supply chains and practices to be carried out for low-carbon construction.

The chapter examined the carbon emissions of the construction supply chains in three sequential phases: cradle-to-gate, gate-to-site, and on-site. The cradle-to-gate system boundary, which includes raw material extraction and product manufacturing, contributes significantly to the overall carbon emissions due to the massive consumption of fossil fuels. The gate-to-site phase focuses on transport and logistics management reasons for extensive carbon emissions due to the daily delivery of large quantities of materials to construction sites. The on-site stage includes a diverse range of construction and related activities responsible for substantial emissions.

The findings of this study highlighted several key practices for reducing carbon emissions across all three stages. Utilising renewable energy was identified as a major solution to replace fossil fuel consumption throughout the supply chain, significantly decreasing CO_2 emissions. Additionally, the study emphasised the importance of energy-efficient machinery, regular maintenance to ensure optimal performance, and reusing and recycling waste, particularly within the cradle-to-gate and on-site system boundaries. For instance, recycled aluminium can save up to 95% of emissions compared to primary aluminium production. Moreover, off-site construction results in an average of 15.6% embodied carbon reduction compared to traditional buildings. In the gate-to-site stage, practices such as local sourcing, intermodal transportation and following policies, standards, and regulations were found to be vital in moving towards a low-carbon construction industry. For example, adopting policies such as construction logistics setup (CLS) can significantly decrease CO_2 emissions from construction transportation by up to 30%.

This chapter contributes to the body of knowledge as a blueprint for both researchers and practitioners. Researchers can use this as a foundation to explore low-carbon methods and technologies and conduct case studies to distinguish the impact of various practices. Then, practitioners can use this as a guide to minimise carbon emissions throughout their operations. Future research may focus on developing low-carbon materials, assessing the carbon footprint of off-site construction, and exploring the role of technology adoption in reducing carbon emissions in the construction sector.

References

1. Abdelaziz EA, Saidur R, Mekhilef S (2011) A review on energy saving strategies in industrial sector. Renew Sustain Energy Rev 15(1):150–168. https://doi.org/10.1016/j.rser.2010.09.003
2. Akbarnezhad A, Xiao J (2017) Estimation and minimization of embodied carbon of buildings: a review. Buildings 7(1). https://doi.org/10.3390/buildings7010005
3. Alsheyab MAT (2021) Recycling of construction and demolition waste and its impact on climate change and sustainable development. Int J Environ Sci Technol 19(3):2129–2138. https://doi.org/10.1007/s13762-021-03217-1
4. Althoey F, Ansari WS, Sufian M, Deifalla AF (2023) Advancements in low-carbon concrete as a construction material for the sustainable built environment. In: Developments in the built environment, p 100284
5. Anuradha G, Perera K, Ekanayake B (2022) Significance of whole life embodied energy and embodied carbon of wall materials compared to their initial/maintenance costs. Int J Constr Manag 22(6):1090–1103
6. Arogundade S, Dulaimi M, Ajayi S (2023) Holistic review of construction process carbon-reduction measures: a systematic literature review approach. Buildings 13(7). https://doi.org/10.3390/buildings13071780

7. Ashworth A, Perera S (2018) Contractual procedures in the construction industry. Routledge, New York
8. Biswas N, Saha P, Mitra S, Mitra S, Majumdar G (2023) Green building material (GBM) vs conventional building material (CBM) to reduce environmental impact
9. Bragin MA, Ye Z, Yu N (2024) Toward efficient transportation electrification of heavy-duty trucks: joint scheduling of truck routing and charging. Transp Res Part C Emerging Technol 160. https://doi.org/10.1016/j.trc.2024.104494
10. Bui TTP, Wilkinson S, Domingo N, MacGregor C (2021) Zero carbon building practices in Aotearoa New Zealand. Energies 14(15):4455
11. Chang B, Kendall A (2011) Life cycle greenhouse gas assessment of infrastructure construction for California's high-speed rail system. Transp Res Part D Transp Environ 16(6):429–434
12. Chang CC, Shi W, Mehta P, Dauwels J (2019) Life cycle energy assessment of university buildings in tropical climate. J Clean Prod 239:117930
13. Chen S, Teng Y, Zhang Y, Leung CKY, Pan W (2023) Reducing embodied carbon in concrete materials: a state-of-the-art review. Resour Conserv Recycl 188. https://doi.org/10.1016/j.resconrec.2022.106653
14. Cheng EWL, Li H, Love PE, Irani Z (2001) An e-business model to support supply chain activities in construction. Logistics Inf Manage 14(1/2):68–78
15. Christopher M, Ryals LJ (2014) The supply chain becomes the demand chain. J Bus Logist 35(1):29–35
16. Costa F, Ribeiro D (2020) Reduction in CO_2 emissions during production of cement, with partial replacement of traditional raw materials by civil construction waste (CCW). J Clean Prod 276:123302
17. Crishna N, Banfill P, Goodsir S (2011) Embodied energy and CO_2 in UK dimension stone. Resour Conserv Recycl 55(12):1265–1273
18. Dsilva J, Zarmukhambetova S, Locke J (2023) Assessment of building materials in the construction sector: a case study using life cycle assessment approach to achieve the circular economy. Heliyon 9(10)
19. Fang Z, Yan J, Lu Q, Chen L, Yang P, Tang J et al (2023) A systematic literature review of carbon footprint decision-making approaches for infrastructure and building projects. Appl Energy 335. https://doi.org/10.1016/j.apenergy.2023.120768
20. Foster G (2020) Circular economy strategies for adaptive reuse of cultural heritage buildings to reduce environmental impacts. Resour Conserv Recycl 152:104507
21. Gan VJ, Cheng JC, Lo IM (2019) A comprehensive approach to mitigation of embodied carbon in reinforced concrete buildings. J Clean Prod 229:582–597
22. Gautam M, Pandey B, Agrawal M (2018) Carbon footprint of aluminum production: emissions and mitigation. In: Environmental carbon footprints. Elsevier, pp 197–228
23. Giesekam J, Barrett JR, Taylor P (2015) Construction sector views on low carbon building materials. Build Res Inf 44(4):423–444. https://doi.org/10.1080/09613218.2016.1086872
24. Green Construction Board (2015) Route map for the built environment
25. Guerlain C, Renault S, Ferrero F (2019) Understanding construction logistics in urban areas and lowering its environmental impact: a focus on construction consolidation centres. Sustainability 11(21):6118
26. Gupta, Dutta (2011) Modeling of financial supply chain. Eur J Oper Res 211(1):47–56. https://doi.org/10.1016/j.ejor.2010.11.005
27. Hu M (2023) Strategies and techniques of life cycle–embodied carbon reduction from the building and construction sector: a review. J Archit Eng 29(3). https://doi.org/10.1061/jaeied.Aeeng-1591
28. Kabirifar K, Mojtahedi M, Wang C, Tamey VWY (2020) Construction and demolition waste management contributing factors coupled with reduce, reuse, and recycle strategies for effective waste management: a review. J Cleaner Prod 263. https://doi.org/10.1016/j.jclepro.2020.121265
29. Karlsson I, Rootzén J, Johnsson F (2020) Reaching net-zero carbon emissions in construction supply chains—analysis of a Swedish road construction project. Renew Sustain Energy Rev 120. https://doi.org/10.1016/j.rser.2019.109651

30. Khan M, Lao J, Dai J-G (2022) Comparative study of advanced computational techniques for estimating the compressive strength of UHPC. J Asian Concr Federation 8(1):51–68
31. Kim B, Lee H, Park H, Kim H (2012) Greenhouse gas emissions from onsite equipment usage in road construction. J Constr Eng Manag 138(8):982–990. https://doi.org/10.1061/(asce)co.1943-7862.0000515
32. Lawrence A, Thollander P, Andrei M, Karlsson M (2019) Specific Energy Consumption/use (SEC) in energy management for improving energy efficiency in industry: meaning, usage and differences. Energies 12(2). https://doi.org/10.3390/en12020247
33. Li Z, Luan W, Zhang Z, Su M (2020) Relationship between urban construction land expansion and population/economic growth in Liaoning Province, China. Land Use Policy 99:105022
34. Lundesjö G (2018) Consolidation centres construction logistics. In: Urban logistics: management, policy and innovation in a rapidly changing environment, vol 210
35. Luo L, He Q, Jaselskis EJ, Xie J (2017) Construction project complexity: research trends and implications. J Constr Eng Manage 143(7). https://doi.org/10.1061/(asce)co.1943-7862.0001306
36. Machacek E, Richter J, Lane R (2017) Governance and risk–value constructions in closing loops of rare earth elements in global value chains. Resources 6(4). https://doi.org/10.3390/resources6040059
37. Makarigakis A, Jimenez-Cisneros B (2019) UNESCO's contribution to face global water challenges. Water 11(2). https://doi.org/10.3390/w11020388
38. Mannan M, Al-Ghamdi SG (2020) Environmental impact of water-use in buildings: latest developments from a life-cycle assessment perspective. J Environ Manage 261:110198. https://doi.org/10.1016/j.jenvman.2020.110198. https://www.ncbi.nlm.nih.gov/pubmed/32148271
39. Martek I, Chen C (2016) Value chain supply procurement strategies in international construction. Manag Decis 54(2):501–521. https://doi.org/10.1108/md-01-2015-0018
40. Minunno R, O'Grady T, Morrison GM, Gruner RL (2021) Investigating the embodied energy and carbon of buildings: a systematic literature review and meta-analysis of life cycle assessments. Renew Sustain Energy Rev 143. https://doi.org/10.1016/j.rser.2021.110935
41. Nanayakkara S, Perera S, Senaratne S, Weerasuriya GT, Bandara HMND (2021) Blockchain and smart contracts: a solution for payment issues in construction supply chains. Informatics 8(2). https://doi.org/10.3390/informatics8020036
42. Nassar R, Saeed D, Sufyan-Ud-Din M, Nassar S (2022) Production of eco-friendly concrete masonry units using powder waste glass. Civil Eng Archit 10:415–424
43. Noke H, Hughes M (2010) Climbing the value chain. Int J Oper Prod Manag 30(2):132–154. https://doi.org/10.1108/01443571011018680
44. Pandit JK, Watson M, Qader A (2020) Reduction of greenhouse gas emissions in steel production. CO2CRC Ltd, Melbourne, Australia
45. Papadopoulos G, Zamer N, Gayialis SP, Tatsiopoulos IP (2016) Supply chain improvement in construction industry. Univ J Manage 4(10):528–534. https://doi.org/10.13189/ujm.2016.041002
46. Passer A, Lasvaux S, Allacker K, De Lathauwer D, Spirinckx C, Wittstock B et al (2015) Environmental product declarations entering the building sector: critical reflections based on 5 to 10 years' experience in different European countries. Int J Life Cycle Assess 20:1199–1212
47. Paulsen JS, Sposto RM (2013) A life cycle energy analysis of social housing in Brazil: case study for the program "MY HOUSE MY LIFE." Energy Build 57:95–102
48. Perera S, Gamage A (2023) Embodied carbon estimating: moving from static to dynamic datasets. In: Built environment economist: Australia and New Zealand, pp 8–11
49. Perera S, Nanayakkara S, Rodrigo M, Senaratne S, Weinand R (2020) Blockchain technology: is it hype or real in the construction industry? J Ind Inf Integr 17:100125
50. Petrovic B, Myhren JA, Zhang X, Wallhagen M, Eriksson O (2019) Life cycle assessment of a wooden single-family house in Sweden. Appl Energy 251:113253
51. Pomponi F, Moncaster A (2016) Embodied carbon mitigation and reduction in the built environment—what does the evidence say? J Environ Manage 181:687–700

52. Raza A, Gholami R, Rezaee R, Rasouli V, Rabiei M (2019) Significant aspects of carbon capture and storage—a review. Petroleum 5(4):335–340. https://doi.org/10.1016/j.petlm.2018.12.007
53. Rodrigo M, Perera S, Senaratne S, Jin X (2019) Embodied carbon mitigation strategies in the construction industry. In: CIB world building congress, pp 1–10
54. Seo M-S, Kim T, Hong G, Kim H (2016) On-site measurements of CO_2 emissions during the construction phase of a building complex. Energies 9(8). https://doi.org/10.3390/en9080599
55. Sezer AA, Fredriksson A (2021) Environmental impact of construction transport and the effects of building certification schemes. Resour Conserv Recycl 172. https://doi.org/10.1016/j.resconrec.2021.105688
56. Singh G, Anil Kumar TC, Naikan VNA (2019) Efficiency monitoring as a strategy for cost effective maintenance of induction motors for minimising carbon emission and energy consumption. Reliab Eng Syst Saf 184:193–201. https://doi.org/10.1016/j.ress.2018.02.015
57. Sizirici B, Fseha Y, Cho C-S, Yildiz I, Byon Y-J (2021) A review of carbon footprint reduction in construction industry, from design to operation. Materials 14(20):6094. https://mdpi-res.com/d_attachment/materials/materials-14-06094/article_deploy/materials-14-06094.pdf?version=1634285012
58. Sundquist V, Gadde L-E, Hulthén K (2018) Reorganising construction logistics for improved performance. Constr Manag Econ 36(1):49–65
59. Tangsucheeva R, Prabhu V (2014) Stochastic financial analytics for cash flow forecasting. Int J Prod Econ 158:65–76. https://doi.org/10.1016/j.ijpe.2014.07.019
60. Teng Y, Li K, Pan W, Ng T (2018) Reducing building life cycle carbon emissions through prefabrication: evidence from and gaps in empirical studies. Build Environ 132:125–136
61. Tien NH, Anh DBH, Thuc TD (2019) Global supply chain and logistics management. Academic Publications, Dehli
62. Valentini L (2023) Sustainable sourcing of raw materials for the built environment. In: Materials today: proceedings
63. Wang Y, Fang X, Yin S, Chen W (2021) Low-carbon development quality of cities in China: evaluation and obstacle analysis. Sustain Cities Soc 64:102553
64. Wu P, Song Y, Zhu J, Chang R (2019) Analysing the influence factors of the carbon emissions from China's building and construction industry from 2000 to 2015. J Clean Prod 221:552–566. https://doi.org/10.1016/j.jclepro.2019.02.200
65. Xin L, Sun H, Xia X, Wang H, Xiao H, Yan X (2022) How does renewable energy technology innovation affect manufacturing carbon intensity in China? Environ Sci Pollution Res 29(39):59784–59801 (Medline, NLM). https://doi.org/10.1007/s11356-022-20012-8. https://www.ncbi.nlm.nih.gov/pubmed/35394630
66. Yakovleva N, Nickless E (2022) Routledge handbook of the extractive industries and sustainable development. Routledge, Abingdon, UK
67. Yang X, Hu M, Wu J, Zhao B (2018) Building-information-modeling enabled life cycle assessment, a case study on carbon footprint accounting for a residential building in China. J Clean Prod 183:729–743
68. Zhang X, Jiao K, Zhang J, Guo Z (2021) A review on low carbon emissions projects of steel industry in the world. J Cleaner Prod 306. https://doi.org/10.1016/j.jclepro.2021.127259
69. Zhao S, Wang H-M, Chen W-Q, Yang D, Liu J-R, Shi F (2019) Environmental impacts of domestic resource extraction in China. Ecosystem Health Sustain 5(1):67–78

Maintenance and Refurbishment for Low Emissions in Buildings

Joseph H. K. Lai

Abstract As climate change poses significant challenges globally, the built environment's contribution to greenhouse gas emissions is increasingly scrutinized. This chapter emphasizes the importance of effective building maintenance and refurbishment strategies to minimize environmental impact. In addition to exploring various maintenance strategies for reducing energy consumption and emissions, the chapter discusses the significance of building refurbishment and smart retrofits for sustainability. An examination of greenhouse gas emissions associated with the built environment is included, along with quantification and reporting methods based on the Greenhouse Gas Protocol. Further to addressing performance benchmarking on carbon emissions among buildings, the chapter outlines methods for conducting environmental and economic evaluations of energy retrofits, showcasing a credible procedure for assessing environmental benefits and related costs. Finally, an empirical case study illustrates practical applications of these evaluations, offering valuable insights for practitioners, decision-makers, and researchers striving for a more sustainable built environment.

1 Introduction

As the global community continues to grapple with the pressing challenge of climate change, the role of the built environment in mitigating greenhouse gas emissions has come under increasing scrutiny. Existing buildings account for a significant portion of worldwide carbon emissions, with the energy required for their operation, maintenance, and refurbishment being a major contributor. This has placed a renewed emphasis on developing effective strategies for maintaining and retrofitting buildings in a way that minimizes their environmental impact.

Building maintenance and refurbishment practices play a critical role in reducing the carbon footprint of the built infrastructure. Proper and proactive maintenance

J. H. K. Lai (✉)
Department of Building Environment and Energy Engineering, The Hong Kong Polytechnic University, Hong Kong, China
e-mail: joseph.lai@polyu.edu.hk

N. Domingo et al. (eds.), *Global Net Zero Carbon Practices in Construction*, Green Energy and Technology, https://doi.org/10.1007/978-981-96-3853-6_8

can extend the lifespan of buildings, improve their energy efficiency, and reduce the need for more resource-intensive renovations. Additionally, energy-efficient retrofits and smart building technologies can significantly lower the operational emissions of existing structures. Understanding the interplay between building maintenance, refurbishment, and emissions reduction is essential for building owners, facility managers, and policymakers to make informed decisions and create a more sustainable built environment.

This chapter will delve into the multifaceted aspects of maintenance and refurbishment for low-emission buildings. It will first explore the importance of building maintenance, examining various maintenance strategies and their impact on energy consumption and emissions. The chapter will then dive into the realm of building refurbishment, discussing the purposes of retrofits, the role of energy-efficient upgrades, and the potential of smart building technologies to enhance sustainability.

Furthermore, the chapter will examine the greenhouse gas emissions associated with the built environment and explains the quantification and reporting method according to the widely recognized Greenhouse Gas Protocol. Performance benchmarking will also be addressed, providing insights into the evaluation and comparison of carbon emissions between existing buildings.

Finally, the chapter will illustrate how to conduct environmental and economic evaluations of building energy retrofits. It will show a credible procedure for assessing the environmental benefits of implementing retrofits and the associated costs that influence investment decisions. An empirical case study will be presented to demonstrate the practical application of these evaluations, offering valuable insights for practitioners, decision-makers as well as researchers in this area.

2 Methodology

A number of steps were taken to prepare this chapter. Firstly, reputable and reliable sources of the most relevant and authoritative sources of information that are germane to the topic of this chapter were identified. Such sources include academic journal articles retrieved from literature databases such as Scopus and ScienceDirect, guidebooks published by professional institutions, textbooks, reports, and credible online resources. In searching such materials, keywords including "carbon", "maintenance", "refurbishment", and its synonym "retrofit" were used in combinations.

Secondly, an extensive review of the literature and online materials identified from the above step was conducted in order to understand the contemporary state of knowledge of the present topic and identify salient information for incorporation into this chapter. Thirdly, detailed notes were taken when reading through the identified sources. Facilitated by a structured note-taking system, this step recorded the bibliographic information for each source, enabling ready retrieval of the information and proper citation of the references in a later step. Fourthly, key data and findings from the past relevant studies were identified from the sources for inclusion in the writing for this chapter.

Fifthly, the various pieces of information gathered from the above steps were collated. This step, which is the most challenging and time-consuming, synthesized information for consolidation into this chapter. To ensure that the information is structured in a logical way with smooth transitioning between sections, related ideas contained in the information were grouped together. After several rounds of editing and proofreading, eventually, the present chapter was finalized.

3 Building Maintenance

3.1 Importance of Maintenance

Maintaining buildings is crucial for ensuring their long-term sustainability and reducing their environmental impact. Proper maintenance can extend the lifespan of a building, improve its energy efficiency, and reduce the need for costly and carbon-intensive refurbishments or replacements [3, 16, 20].

One of the primary benefits of regular building maintenance is the reduction in energy consumption. Well-maintained building systems, such as heating, ventilation, and air-conditioning (HVAC), lighting, and insulation, operate more efficiently, leading to lower energy usage and associated greenhouse gas emissions. By addressing issues like air leaks, malfunctioning equipment, and suboptimal settings, maintenance can significantly improve a building's overall energy performance.

Furthermore, maintenance helps preserve the integrity of a building's structure and materials. This is particularly important for older buildings, where deterioration can compromise the building envelope and increase the risk of air and moisture infiltration. Timely repairs and preventive measures can prevent more extensive and carbon-intensive renovations in the future.

Regular maintenance also contributes to the overall comfort and health of building occupants. Well-functioning systems, such as ventilation and temperature control, ensure that indoor air quality is maintained, which can have a positive impact on the health and well-being of the occupants. Additionally, maintaining building systems can help reduce the risk of equipment failures and associated disruptions, ensuring a more reliable and comfortable environment for the building's users.

3.2 Maintenance Strategies

Effective building maintenance involves the implementation of comprehensive strategies that address both preventive and reactive measures. Preventive maintenance focuses on regular inspections, cleaning, and routine upkeep to identify and address issues before they escalate [6]. This can include tasks such as

- checking and servicing HVAC systems to ensure optimal efficiency and air quality;

- inspecting and maintaining the building envelope, including windows, doors, and insulation, to minimize air leaks;
- cleaning and calibrating building automation systems to ensure they are operating as intended;
- performing regular maintenance on lighting fixtures and controls to maintain energy efficiency;
- inspecting and maintaining the building's electrical and plumbing systems to prevent issues and ensure efficient operation;
- regularly cleaning and inspecting the building's exterior, including the roof, gutters, and landscaping, to prevent water damage and other issues.

Reactive maintenance, on the other hand, involves addressing issues that arise unexpectedly, such as equipment breakdowns or unexpected system failures. Swift response and efficient troubleshooting can minimize the impact of these events and prevent them from escalating into larger problems.

Effective maintenance strategies also consider the integration of building management systems (BMS) and smart technologies. These systems can provide real-time data on building performance, allowing facility managers to make informed decisions and prioritize maintenance tasks based on their impact on energy consumption and emissions. By leveraging data analytics and predictive maintenance algorithms, building owners can anticipate and address issues before they occur, further improving the efficiency and sustainability of their properties.

To ensure the effectiveness of maintenance strategies, it is essential to develop and implement a comprehensive maintenance plan that aligns with the building's specific needs and characteristics [5]. This plan should include a schedule of regular inspections, maintenance tasks, and record-keeping to track the building's performance and identify areas for improvement over time.

By implementing a comprehensive maintenance program that combines preventive and reactive measures, building owners and managers can ensure the long-term sustainability of their properties, reduce their environmental impact, and maintain the comfort and well-being of their occupants. Regular maintenance not only helps preserve the building's physical integrity but also contributes to its overall energy efficiency and environmental performance, making it a crucial component of a low-emission building strategy.

4 Building Refurbishment

4.1 Purposes of Refurbishments/Retrofits

Building refurbishments and retrofits play a crucial role in the transition towards low-emission buildings. These interventions go beyond routine maintenance and aim to significantly improve the building's performance, enhance its energy efficiency, and reduce its environmental impact [13].

One of the primary purposes of building refurbishments is to address the issue of aging infrastructure and deteriorating building components. As buildings age, their systems and materials can become less efficient, compromising the overall performance and increasing energy consumption. Refurbishments provide an opportunity to upgrade and replace outdated technologies with more energy-efficient and sustainable alternatives, contributing to long-term emissions reduction.

Another key purpose of building refurbishments is to adapt to changing occupancy patterns, user requirements, and evolving energy standards. As the needs and expectations of building occupants evolve, refurbishments allow building owners to modify and modernize the property to better suit these needs. This can include adjustments to the building layout, improvements to indoor air quality, and the incorporation of smart technologies that enhance energy management and user experience.

Refurbishments can also be driven by the desire to improve the overall environmental footprint of a building. By addressing issues such as poor insulation, inefficient heating and cooling systems, and outdated lighting, refurbishments can significantly reduce a building's energy consumption and associated greenhouse gas emissions. This aligns with the broader goal of creating a more sustainable built environment and mitigating the impact of the construction and real estate sectors on climate change.

4.2 Energy-Efficient Retrofits

One of the most impactful types of building refurbishments is the implementation of energy-efficient retrofits. These interventions focus on improving the building's thermal performance, reducing energy demand, and increasing the efficiency of building systems [11] (Tomrukcu and Ashrafian 2024).

Thermal efficiency upgrades, such as enhancing the building envelope through improved insulation, air sealing, and high-performance windows, can significantly reduce heat loss and gain, leading to decreased heating and cooling requirements. These measures not only lower energy consumption but also improve occupant comfort and well-being.

Replacement of outdated HVAC systems with more efficient, low-emission alternatives is another crucial component of energy-efficient retrofits. This can include the installation of heat pumps, high-efficiency boilers, and advanced air-conditioning systems that utilize eco-friendly refrigerants. These upgrades can dramatically reduce the building's energy consumption and associated greenhouse gas emissions.

Lighting upgrades, including the replacement of traditional bulbs with energy-efficient LED technology, can also contribute to significant energy savings. Integrating lighting controls, daylight harvesting, and occupancy sensors further enhances the building's energy performance and reduces unnecessary lighting usage.

To maximize the impact of energy-efficient retrofits, it is essential to take a holistic approach that considers the interactions between different building systems. By optimizing the integration of these systems, building owners can achieve greater energy savings and emission reductions.

4.3 Smart Retrofits

In addition to energy-efficient retrofits, the integration of smart technologies and building automation systems can significantly enhance the overall performance and sustainability of buildings.

Smart retrofits involve the installation of advanced sensors, controls, and software systems that enable real-time monitoring, optimization, and remote management of building operations. (Peiris et al., 2023). Examples include sensor-based intelligent building management system (BMS) for better ventilation and energy performance by auto-sensed indoor CO_2 concentration and temperature [7] and innovative façade modules consisting of adaptable smart modular heat recovery units [2]. These systems can collect and analyze data on energy consumption, indoor air quality, occupancy patterns, and equipment performance, providing building managers with valuable insights to inform decision-making and improve building efficiency.

One of the key benefits of smart retrofits is the ability to optimize the performance of building systems, leading to reduced energy usage and emissions. Smart building management systems can automatically adjust HVAC, lighting, and other systems based on factors such as occupancy, weather conditions, and energy prices, ensuring that energy is used efficiently and minimizing waste.

Smart retrofits can also enhance the user experience and well-being of building occupants. Intelligent control systems can adjust indoor temperature, humidity, and air quality to maintain optimal comfort levels, while also monitoring and addressing potential health and safety concerns, such as air quality issues or equipment malfunctions.

Furthermore, the integration of renewable energy sources, such as rooftop solar panels, can be facilitated through smart retrofits. These systems can monitor the building's energy generation and consumption, enabling building owners to optimize the use of on-site renewable energy and potentially export excess generation to the grid.

To effectively implement smart retrofits, it is essential to ensure the seamless integration of various building systems and technologies. This may involve the replacement or upgrade of legacy systems, as well as the implementation of open communication protocols and data exchange platforms. Building owners should also consider the cyber-security implications of smart building technologies and implement appropriate measures to protect against potential cyber threats.

By incorporating both energy-efficient and smart retrofits, building owners can achieve significant reductions in energy consumption and greenhouse gas emissions, while also improving the overall functionality, comfort, and sustainability of their properties. These refurbishment strategies are crucial for transforming the existing building stock into a more environmentally responsible and resilient infrastructure, contributing to the broader goals of a low-carbon built environment.

4.4 Carbon Emissions of Buildings

Buildings account for a significant portion of global carbon emissions, contributing around 40% of total global carbon emissions. The major types of buildings that contribute the most to carbon emissions are residential and commercial buildings [18].

Residential buildings, such as houses and apartments, account for a large share of building-related emissions due to the widespread use of fossil fuels for heating and the high energy demands of household appliances. Commercial buildings, including offices, retail spaces, and institutional buildings, also contribute significantly to carbon emissions through their energy-intensive operations and energy consumption required for the operation of buildings, including heating, cooling, lighting, and the use of electrical appliances [19].

Addressing the carbon footprint of buildings is crucial in the transition towards a low-emission future, and strategies such as energy-efficient design, the use of renewable energy sources, and the retrofit of existing buildings can play a vital role in reducing the environmental impact of the built environment.

4.5 The Greenhouse Gas Protocol

Understanding the greenhouse gas (GHG) emissions associated with buildings is a crucial step in developing effective strategies for low-emission maintenance and refurbishment. The Greenhouse Gas Protocol, a widely adopted international standard, provides a comprehensive framework for categorizing and accounting for these emissions [12]. It defines three distinct scopes of emissions as shown in Table 1.

Considering all three scopes of emissions is crucial for building owners and managers to develop a comprehensive understanding of the building's carbon footprint and identify opportunities for emissions reduction. While Scope 1 and Scope 2 emissions are more directly within the control of the building owner, Scope 3 emissions can also be influenced through strategic procurement, waste management, and collaboration with supply chain partners.

Table 1 Three distinct scopes of carbon emissions

Scope 1: Direct emissions	Scope 1 emissions are those that arise directly from sources owned or controlled by the building. This includes the combustion of fuels for heating, cooling, and on-site power generation, as well as any other direct sources of GHG emissions within the building's boundaries
Scope 2: Indirect emissions from energy use	Scope 2 emissions are indirect emissions resulting from the generation of purchased electricity, heat, steam, or cooling consumed by the building. These emissions occur outside the building's boundaries but are a consequence of the building's energy use
Scope 3: Other indirect emissions	Scope 3 emissions are all other indirect emissions that occur in the building's value chain, both upstream and downstream. This includes emissions from the extraction and production of purchased materials and fuels, transportation of occupants and goods, waste disposal, and even the end-of-life treatment of the building itself

4.6 Quantification and Reporting

Accurately quantifying and reporting the GHG emissions associated with a building is a crucial step in developing effective low-emission strategies. This process, often referred to as a carbon audit [17], involves gathering and analyzing data from various sources to determine the building's total carbon footprint.

For Scope 1 emissions, building owners can rely on metered data or invoices for fuel consumption, such as natural gas, heating oil, or propane, and then apply appropriate emission factors to convert these inputs into GHG emissions. Scope 2 emissions are typically calculated based on electricity consumption data and the emission factors associated with the local or regional electricity grid.

Quantifying Scope 3 emissions can be more challenging, as it requires collecting data from a broader range of sources, including supply chain partners, waste management providers, and transportation records. Building owners may need to make reasonable assumptions or use industry-specific data to estimate these indirect emissions.

Once the GHG emissions have been quantified, the next step is to establish a system for regular reporting and monitoring. This can involve the creation of an annual sustainability report or the integration of emissions data into the building's overall performance management system. Consistent and transparent reporting not only helps building owners track their progress over time but also supports external benchmarking and comparison with industry peers.

The process of quantification and reporting can be further enhanced through the use of specialized software tools and data management platforms. These solutions can streamline the data collection, analysis, and reporting processes, making it easier for building owners to maintain accurate and up-to-date records of their carbon emissions.

One example of such a tool is the Carbon Risk Real Estate Monitor (CRREM), a free online platform developed by a consortium of European research institutions and industry partners [4]. CRREM provides building owners with a comprehensive assessment of their GHG emissions, energy performance, and alignment with global climate targets, such as the Paris Agreement. By integrating CRREM into their emissions management strategy, building owners can make informed decisions about maintenance and refurbishment priorities, identify cost-effective emissions reduction measures, and track their progress towards decarbonization goals.

4.7 *Performance Benchmarking*

Performance benchmarking is a valuable tool for building owners and managers to assess the carbon emissions performance of their buildings and identify areas for improvement. By comparing their building's emissions against industry benchmarks or similar building types, they can gain valuable insights and make informed decisions regarding maintenance and refurbishment strategies.

There are several established benchmarking frameworks and databases that building owners can utilize to assess their performance. One prominent example is the Energy Star Portfolio Manager [9], a free online tool developed by the U.S. Environmental Protection Agency (EPA) that allows users to track and benchmark the energy and emissions performance of their buildings.

Another widely used benchmarking resource is the Building Energy Quotient (Building EQ) program [1], which provides a standardized approach to evaluating a building's energy use and emissions. The Building EQ rating system considers various factors, including building type, size, occupancy, and climate, to establish a comprehensive performance assessment.

At the global level, initiatives such as the World Green Building Council's "Advancing Net Zero" program and the International Energy Agency's "Energy in Buildings and Communities" program offer robust benchmarking frameworks and databases that can help building owners compare their performance against international standards and best practices [8] (World Green Building Council 2024).

By engaging in performance benchmarking, building owners can

1. identify areas of high emissions and prioritize maintenance and refurbishment efforts accordingly;
2. track their progress over time and measure the impact of implemented low-emission strategies;
3. learn from the best practices of high-performing buildings and adapt these strategies to their own properties;
4. demonstrate their commitment to sustainability and emissions reduction to stakeholders, including tenants, investors, and regulatory bodies.

Performance benchmarking not only supports informed decision-making but also fosters a culture of continuous improvement, driving building owners to strive for ever-higher levels of carbon emissions reduction and environmental stewardship.

One notable example of a building that has achieved exemplary emissions performance is the Empire State Building in New York City. Through a comprehensive retrofit program, the building's owners were able to reduce the building's energy use by 38% and its carbon emissions by 40%, making it one of the most energy-efficient skyscrapers in the world [14]. By sharing their experiences and best practices, the Empire State Building's owners have inspired other building owners to emulate their success and work towards a more sustainable built environment.

In addition to established benchmarking frameworks, building owners can also explore emerging technologies and data-driven approaches to assess and optimize their emissions performance. For instance, the use of sensor networks, machine learning algorithms, and digital building information models (BIMs) can provide real-time insights into a building's energy use and emissions, enabling more targeted and responsive maintenance and refurbishment strategies.

As the global community continues to prioritize the reduction of greenhouse gas emissions, the role of building owners and managers in driving this transition becomes increasingly critical. By leveraging appropriate tools and frameworks for carbon mitigation, building owners can not only improve the environmental performance of their properties but also contribute to the broader societal goal of mitigating climate change and creating a more sustainable built environment.

4.8 Environmental and Economic Evaluations of Building Energy Retrofits

Performing robust environmental and economic analyses is crucial when considering building energy retrofits. Such evaluations help quantify the potential energy savings, emissions reductions, and financial impacts of efficiency upgrades. However, these assessments can be challenging in practice. Changes in weather conditions, occupant behavior, and future energy prices are among the parameters that are difficult to predict accurately. Still, these evaluations are essential for justifying retrofit investments and ensuring projects deliver the expected environmental and economic benefits.

5 Environmental Evaluation

The environmental performance of an energy-saving measure (ESM) can be evaluated by measuring the resulting reduction in carbon emissions. Following the guidelines issued by the Environmental Protection Department and the Electrical and Mechanical Services Department [10], which was modelled on the Greenhouse Gas (GHG) Protocol [21], the first step is to quantify the three different scopes of carbon emissions associated with the building's energy consumption.

Referencing the calculation procedure outlined by Lai [15], the first component of carbon emission, due to on-site fuel combustion (under scope 1 of the GHG Protocol) and denoted as CE_A^D (in tons of CO_2-equivalent), can be calculated by Eq. (1):

$$CE_A^D = \sum_{f=1}^{f=F} \sum_{t=1}^{t=T} A_{f,t} \cdot F_{f,A} \cdot G_A \quad (1)$$

The second component of carbon emission, due to consumption of purchased electricity (under scope 2 of the GHG Protocol), i.e. CE_{GHG}^E, can be calculated by Eq. (2):

$$CE_{GHG}^E = \sum_{t=1}^{t=T} A_{E,t} \cdot F_{,t} \quad (2)$$

The third component of carbon emission (under scope 3 of the GHG Protocol), which can be calculated by Eq. (3), covers the emission due to electricity used for freshwater supply and processing sewage (i.e., CE_{GHG}^{SS}):

$$CE_{GHG}^{SS} = A_W(F_W + F_D) \quad (3)$$

Based on the amounts of carbon emissions determined from Eqs. (1) to (3), the total carbon emissions, respectively, in the periods before and after the ESM is implemented, can be calculated by Eq. (4), where $CE_x = CE_a$, i.e., carbon emission after ESM implementation (kg CO_2-e), or CE_b, i.e., carbon emission before ESM implementation (kg CO_2-e).

$$CE_x = CE_A^D + CE_{GHG}^E + CE_{GHG}^{SS} \quad (4)$$

Using Eq. (5), the amount of carbon emission reduced from implementing the ESMs (i.e., ΔCR_t) can be calculated:

$$\Delta CR_t = CE_b - CE_a \quad (5)$$

where

$A_{E,t}$ = amount of electricity used in the t-th period (kWh);
$A_{f,t}$ = amount of fuel f consumed in the t-th period;
A_W = amount of water consumed (m^3);
F_D = emission factor of processing sewage (kg CO_2-e/m^3);
$F_{E,t}$ = emission factor of electricity used in the t-th period (kg CO_2-e/kWh);
$F_{f,A}$ = emission factor of gas A (e.g. CO_2, CH_4 or N_2O) for fuel f;
F_W = emission factor of freshwater supply (kg CO_2-e/m^3); and
G_A = global warming potential of gas A.

6 Economic Evaluation

To evaluate the economic performance of the implemented ESMs, two key indicators are useful: (1) return on investment (ROI) and (2) net present value (NPV). ROI reflects the profitability of an investment in an ESM by measuring the ratio of the monetary benefits to the total costs of the ESM. NPV is used to translate future monetary values (e.g., savings in electricity bill, operation and maintenance (O&M) cost of the ESM) into their current equivalents.

Using Eqs. (6) and (7), the NPV and ROI of an ESM can be calculated:

$$NPV_B = \sum_{t=1}^{n} (B_t - C_t)/(1+i)^t \tag{6}$$

$$ROI = (NPV_B - NPV_I)/NPV_I \tag{7}$$

For an investment made on an ESM before the base time at which the economic evaluation refers, its net present value (i.e., NPV_I) can be determined using Eq. (8).

$$NPV_I = \sum_{t=1}^{n*} I/(1+i)^t \tag{8}$$

As different types of equipment (e.g. heat recovery chiller, variable speed pump) may have different functional lifespans, the length of time during which an ESM can contribute to energy saving, subject to which the NPV is calculated, varies from one type of ESM to another. In general, the energy efficiency of an equipment drops over time; for example, the wear and tear of the mechanical parts of a water pump reduces the pump's efficiency. Such a decay in energy saving is referred to as "degradation" of equipment. Thus, a degradation factor (δ) reflects the annual rate according to which the EMS loses its energy-saving capacity. The calculation of δ generally depends on the type and situation of an ESM implemented.

Using Eq. (9), the monetary benefit gained from the implementation of an EMS at a certain time interval can be calculated:

$$B_t = B_1 \cdot \delta_t \tag{9}$$

where

B_1 = monetary benefit gained from the ESM in the first time interval;
B_t = monetary benefit gained between time intervals t and $t-1$;
C_t = O&M cost increase (positive value) or decrease (negative value) resulting from implementing the ESM;
δ_t = degradation factor of the equipment at time interval t;
i = discount rate of money;
n = number of time intervals within the ESM's functional lifespan;

n^* = number of time intervals between the investment was made and the evaluation base time;
NPV_B = net present value of monetary benefit gained from an ESM after cost;
NPV_I = net present value of investment on the ESM; and
t = unit time interval within the evaluation period.

7 Environmental Cum Economic Evaluation

To evaluate the cost-effectiveness of an ESM, it is essential to assess both the environmental and economic performances of the ESM. For this purpose, Zheng and Lai (2018) introduced a metric known as carbon reduction efficiency (*CRE*). It is an indicator (unit: kg/$) that gauges how much the amount of a building's carbon emission can be reduced per unit present value of cost incurred for implementing the ESM. Thus, the CRE of an ESM can be calculating by dividing the amount of reduction in carbon emission by the sum of net present values of the investment cost and change in O&M cost associated with the ESM, i.e., Eq. (10):

$$CRE = \sum_{t=1}^{n} \Delta CR_t / (NPV_I + NPV_C) \quad (10)$$

where
NPV_C = net present value of the change in O&M cost in using the ESM.

8 An Empirical Case Study

Credible, empirical evaluation studies on energy-saving measures (ESMs) are limited, for reasons such as data unavailability. Despite such difficulties, a study was initiated to evaluate the economic and environmental performances of ESMs that had been implemented in buildings. The project team collected detailed data through interviews. This included information on the physical characteristics of the buildings, such as age, number of floors, and total floor area. The team also gathered monthly data on the energy end-uses of the buildings and associated costs of the ESMs over a 5-year period. From the buildings included in the sample, the one with the most comprehensive data available was selected for a case study.

As reported by Zheng and Lai (2018), the case study building is a curtain-walled commercial building located in downtown Hong Kong. The building commenced its operation in 1991 and has a total internal floor area of approximately 21,039 m^2. This is divided between 15 floors of office space (13,051 m^2), 3 floors of retail space (3606 m^2), and 4382 m^2 of common areas. The common spaces include lobbies, stairwells, back-of-house facilities, and shared corridors outside the office and retail tenancies.

The building is served by a central chiller plant located on the roof. This plant consists of four chillers, each with a cooling capacity of 320 tons of refrigeration. The central plant provides air-conditioning (AC) to the office areas, retail spaces, lobbies, and corridors. The air-conditioning terminals installed in the tenant spaces are fan coil units, which allow occupants to adjust the temperature set points to meet their needs.

The primary energy source for operating this building is electricity. In addition to powering the central chiller plant and lighting the common areas, the landlord's electricity consumption also covers communal facilities such as lifts, water pumps, ventilation fans, and fire service installations. The building also has a diesel-powered emergency generator that is used during power outages or periodic functional testing.

8.1 The Retrofits

In an effort to reduce energy use and mitigate carbon emissions, the facility management (FM) team of the building decided to implement several ESMs. Given the hot and humid subtropical climate of Hong Kong, the AC systems in most commercial buildings operate for extended periods each year, consuming a substantial amount of energy. The FM team recognized that retrofitting the building's original air-cooled AC system could lead to significant improvements.

Towards this end, the team installed a freshwater cooling system and added a central control and monitoring system (CCMS) for the chiller plant. An automated cleaning system was also installed to prevent the chiller's heat exchanger tubes from fouling, thereby improving their heat exchange efficiency. Additionally, dedicated sub-meters were installed to track the energy consumption of the retrofitted systems. The implementation of the ESMs was commenced in February of Year 5 (of the 5-year data period) and completed in November of the same year.

8.2 Carbon Emissions

Using Eqs. (1)–(4), the carbon emissions of the building over the five years were computed. The major emission source was the use of electricity purchased from the power company for running the building's electrical installations, and the corresponding amount of carbon emission was as high as 2140.7 tons per year, representing 99.8% of the building's total emission. In contrast, the carbon emissions due to the consumption of electricity for water supply and processing sewage were minimal, and the emissions due to the use of diesel oil for emergency power generation were even negligible.

Figure 1 plots the monthly carbon emissions pertaining to the AC system over the five years. From Year 1 to Year 4, the variations of the carbon emissions under scope

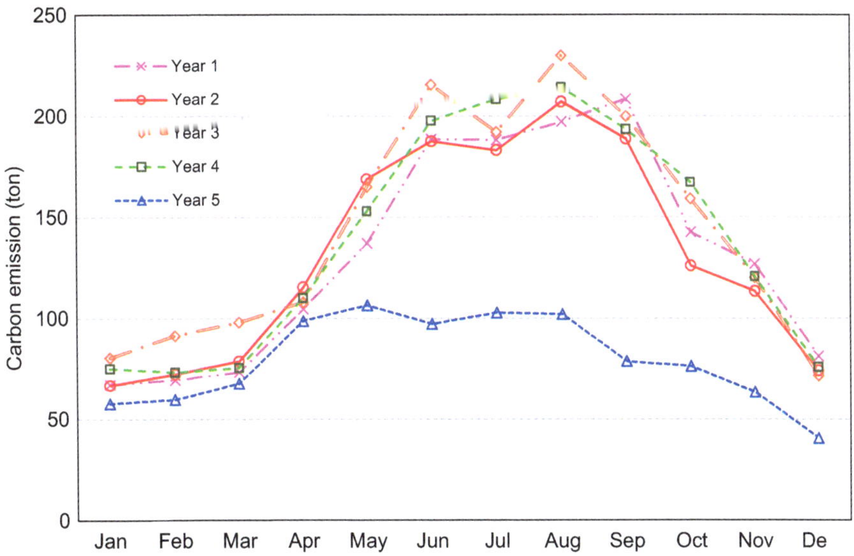

Fig. 1 Monthly carbon emissions

2 of the GHG Protocol were not obvious. In Year 5, the annual amount of scope 2 carbon emission dropped significantly to 1250.7 tons.

Throughout the five-year period, the peak carbon emissions occurred in summer (June to September), while the lowest emission level was found in winter (December). At the beginning of Year 5 where the implementation of the ESM commenced, part of the existing AC system was suspended to facilitate the retrofit work. As a result, the electricity consumption and hence the carbon emission of the system started to decrease.

8.3 Retrofit and Maintenance Costs

The initial costs for implementing the ESMs were as follows:

- Install a freshwater cooling system for the chiller plant: $2,101,076.
- Install an automatic tube cleaning system for chillers: $95,769.
- Install a CCMS for the chiller plant: $184,615.

Even if the plant were not retrofitted with a freshwater cooling system, according to the building's FM team, an expenditure of $1,260,645 was still needed to replace the old chiller plant with a new air-cooled chiller plant, Therefore, the actual additional cost for installing a new water-cooled chiller plant is $840,431 (i.e. $2,101,076–$1,260,645).

Table 2 Carbon reduction and carbon reduction efficiency

Emission factor	Degradation factor (0%)		Degradation factor (1.1%)	
	CR (kg)	*CRE* (kg/$)	*CR* (kg)	*CRE* (kg/$)
0.54	9,823,385	5.998	8,796,841	5.371
0.64	11,642,530	7.109	1,0425,886	6.366

The cost of the retrofits consists of the initial expenditure for implementing the ESMs and the increase in O&M cost for the retrofitted system. This latter additional cost, which is $27,225 per year, stems from the water consumption for the new water-cooled system and services including regular cleaning of the cooling towers, water treatment against Legionella, and sampling and testing of the condenser water quality.

8.4 Environmental-Economic Performance

The amount of carbon emission reduced from implementing the ESMs was calculated using Eq. (5). Afterwards, NPV and ROI of the ESMs were computed by Eqs. (6)–(8). In this calculation process, different discount rates of money (in Eq. (8)) and different degradation factors of the equipment (in Eq. (9)), were considered. Finally, the carbon reduction efficiency of the ESMs was determined using Eq. (10). For illustrative purposes, the calculated values of carbon reduction and carbon reduction efficiency for different scenarios (emission factors (0.54 and 0.64) and degradation factors (0% and 1.1%)) are as summarized in Table 2.

While it is important to ascertain the amount of carbon reduction obtainable from implementing retrofits, carbon reduction efficiency, which reflects the amount of carbon emission reduced per unit price of investment on the retrofit, serves as a useful indicator for decision-makers. This helps (a) building owners to decide on whether to invest on retrofits; (b) building designers or engineers to make justification for retrofit proposals; and (c) facility managers to monitor and assess the actual carbon reduction resulting from implementation of retrofits.

9 Summary

Through taking a number of steps to review and synthesize information retrieved from reputable and reliable sources of the most relevant and authoritative literature in this field, this chapter has examined the multiple facets of maintenance and refurbishment that can contribute to reducing greenhouse gas emissions from existing buildings.

Building maintenance emerged as a crucial factor, as proper and proactive maintenance can extend the lifespan of buildings, improve their energy efficiency, and minimize the need for resource-intensive renovations. Various maintenance strategies, such as preventive, corrective, and predictive approaches, were discussed in terms of their impacts on building energy consumption and carbon emissions.

The chapter then delved into the realm of building refurbishment, highlighting the purposes of retrofits, which include enhancing energy efficiency, incorporating renewable energy sources, and integrating smart building technologies. Energy-efficient retrofits, such as upgrading thermal insulation, windows, lighting, and HVAC systems, were among the effective means of lowering operational emissions. Additionally, the potential of smart building technologies, including automated controls and real-time monitoring for enhancing the sustainability of existing buildings, was discussed.

Furthermore, the chapter examined the quantification and reporting of greenhouse gas emissions associated with the built environment, drawing on the procedure under the widely recognized Greenhouse Gas Protocol. Performance benchmarking was also addressed, providing insights into the evaluation and comparison of carbon emissions across buildings.

Lastly, the chapter illustrated a credible method for conducting environmental and economic evaluations of building energy retrofits. This included assessing the environmental benefits of implementing retrofits, such as reduced carbon emissions, and analyzing the associated costs that influence investment decisions. An empirical case study was presented to demonstrate the practical application of these evaluations, offering valuable insights for practitioners, decision-makers and researchers in this area.

By highlighting these key points, this chapter equips readers with a comprehensive understanding of the considerations, strategies, and methods necessary to make our built environment more sustainable through effective maintenance and refurbishment.

References

1. American Society of Heating, Refrigerating and Air-Conditioning Engineers (2024) Building EQ. https://www.ashrae.org/technical-resources/building-eq. Accessed on 10 Aug 2024
2. Basso P, Mililli M, Herrero FJ, Sanz R, Casaldiga P (2017) E2VENT–design and integration of an adaptable module for residential building renovation. J Façade Des Eng 5(2):7–23
3. Cao N, Sing MCP (2024) Workforce forecasting in the building maintenance and repair work: evaluating machine learning and LSTM models. J Build Eng 95:110125
4. Carbon Risk Real Estate Monitor (CRREM) (2024) Carbon risk real estate monitor. https://www.crrem.eu/about-crrem/. Accessed on 22 July 2024
5. Chanter B, Swallow P (2007) Building maintenance management. Wiley-Blackwell
6. Chartered Institution of Building Services Engineers (2023) Guide M: maintenance engineering and management. Chartered Institution of Building Services Engineers
7. Che WW, Tso CY, Sun L, Ip DYK, Lee H, Chao CYH, Lau AKH (2019) Energy consumption, indoor thermal comfort and air quality in a commercial office with retrofitted heat, ventilation and air conditioning (HVAC) system. Energy Build 201:202–215

8. Energy in Buildings and Communities Programme (EBC) (2024) International Energy Agency's Energy in buildings and communities programme. https://www.iea-ebc.org/. Accessed on 12 Aug 2024
9. Energy Star (2024) Portfolio Manager. https://portfoliomanager.energystar.gov/pm/login. Accessed 22 July 2024
10. Environmental Protection Department and Electrical and Mechanical Services Department (2010) Guidelines to account for and report on greenhouse gas emissions and removals for buildings (Commercial, Residential or Institutional Purposes) in Hong Kong, Hong Kong Government
11. Fan Y, Xia X (2018) Energy-efficiency building retrofit planning for green building compliance. Build Environ 136:312–321
12. Greenhouse Gas Protocol (2024) Standards & guidance. https://ghgprotocol.org/standards-guidance. Accessed on 22 July 2024
13. Grossi F, Ge H, Zmeureanu R, Baba F (2024) Assessing the effectiveness of building retrofits in reducing GHG emissions: a Canadian school case study. J Build Eng 91:109622
14. Jones Lang LaSalle (JLL) (2024) Client story: Empire State Building retrofit cuts 10-year emissions by 40%. https://www.jll.com.hk/en/client-stories/empire-state-building-retrofit-cuts-10-year-emissions-by-40. Accessed 22 July 2024
15. Lai JHK (2015) Carbon footprints of hotels: analysis of three archetypes in Hong Kong. Sustain Cities Soc 14:334–341
16. Lai JHK, Lu M (2024) Carbon emission and maintenance cost of commercial buildings: quantification, analysis and benchmarking. J Cleaner Prod 447:141459
17. Lai JHK, Yik FWH, Man CS (2012) Carbon audit: a literature review and an empirical study on a hotel. Facilities 30(9):417–431
18. Lu M, Lai J (2019) Building energy: a review on consumptions, policies, rating schemes and standards. Energy Proc 158:3633–3638
19. Lu M, Lai J (2020) Review on carbon emissions of commercial buildings. Renew Sustain Energy Rev 119:109545
20. Ng RTH, Lai JHK, Leung OCH, Edwards DJ (2023) Assessing lift maintenance performance of high-rise residential buildings. J Build Eng 68:106202
21. World Resources Institute (WRI) and World Business Council for Sustainable Development (WBCSD) (2004) The Greenhouse Gas protocol: a corporate accounting and reporting standard

Are Australian Construction and Demolition Waste Management Research Efforts Coherent with Policy Targets?

R. A. A. Dilogini, G. Karunasena, N. Udawatta, and C. Liu

Abstract Australia's 2022 National Waste Report states that between 2016 and 2021, the generation of construction and demolition waste (CDW) grew from 23.4 to 29 million tonnes, making up 38% of all waste generated in the nation. The National Waste Policy Action Plan aims to increase the CDW recycling rate to 80% by 2030 to minimize waste as one of the key priorities. In this context, CDW sector-specific endeavours during the recent past are vital understanding to achieve the set policy priorities. Thus, the chapter aims to investigate whether current construction and demolition waste management (CDWM) research efforts in Australia are coherent with policy targets. A systematic literature review was conducted and the study utilized 126 research outputs published in Web of Science and Scopus databases to achieve this. The PRISMA methodology was employed with imposing year restrictions from 2019 to 2023. Additionally, the study referred to the federal and state/territory-level policy documents to review the policy targets concerning CDWM. The findings revealed that research has been conducted on five major themes: management practices, recycling and upcycling, circular economy, stakeholder behaviour, and technology advancements. However, most current practices are not fully aligned with national and state policy targets. To enhance alignment and effectiveness, further research should focus on integrating policy objectives and addressing existing gaps in practice. These findings will provide a valuable assessment of the current state of CDWM research efforts and serve as a roadmap for researchers and industry practitioners to achieve the policy targets.

1 Introduction

In the 2020/21 period, Australia produced around 75.8 million tonnes of waste, with building and demolition materials making up about 25.1 million tonnes of that total (Department of Climate Change, Energy, the Environment and Water [DCCEEW],

R. A. A. Dilogini · G. Karunasena (✉) · N. Udawatta · C. Liu
School of Architecture & Built Environment, Deakin University, Geelong 3220, Australia
e-mail: gayani.karunasena@deakin.edu.au

N. Domingo et al. (eds.), *Global Net Zero Carbon Practices in Construction*, Green Energy and Technology, https://doi.org/10.1007/978-981-96-3853-6_9

2022). The sheer scale and consequences of construction and demolition waste (CDW) are quite concerning [17], capturing growing attention on a global scale [40]. These significant challenges primarily arise from the construction industry's heavy reliance on conventional building methods and a linear economic model, which operates on a "take, make, dispose" principle ([4, 12, 13]; Illankoon and Vithanage 2023).

The rise in construction and demolition waste (CDW) in Australia is largely attributed to rapid urbanization and the expansion of the construction industry, which have together led to an increase in waste generation [33]. Disposing of these materials in landfills presents serious environmental risks, including soil contamination, water pollution, and greenhouse gas emissions, stemming from the breakdown of organic materials and the leaching of chemicals [40]. Additionally, CDW constitutes a significant portion of the waste sent to landfills, highlighting inefficiencies in the current waste management practices.

In Australia, efforts to manage construction and demolition waste (CDW) more sustainably have centred on boosting recycling and resource recovery rates. Nevertheless, despite these initiatives, the country has fallen short of its recycling targets, with the recycling rate for CDW not meeting the ambitious goals outlined by government policies [1]. There is also an increasing awareness of the necessity to move away from a linear economy model towards a circular economy approach, which focuses on maximizing the use of materials through practices like recycling, reuse, and sustainable design [4, 13]. This transition is vital not only for reducing the environmental impacts of waste but also for conserving natural resources and fostering economic growth [2].

Despite extensive research on various aspects of CDWM, a critical gap remains in evaluating whether these efforts align with the policy objectives set by government authorities. Recent reviews have highlighted that while many studies focus on technological advancements and stakeholder behaviour, there is a limited assessment of how well these practices align with national- or state-level waste management policies [37, 43]. Addressing this gap is crucial for ensuring that research efforts contribute effectively to achieving policy goals and improving waste management practices across the construction industry in Australia.

Therefore, this chapter aims to investigate whether current CDWM efforts in Australia are coherent with policy targets. To achieve this aim, the chapter will: (1) examine the research efforts conducted in Australia over the last 5 years, (2) analyse policy targets set by federal and state-level governments, (3) identify gaps in the current literature regarding policy coherence, and (4) provide recommendations for future research and policy development to enhance the alignment of CDWM practices towards sustainability.

2 Literature Review

Construction and demolition waste (CDW) in Australia presents a major environmental and economic challenge, contributing significantly to the nation's total waste stream [10, 32]. The volume of CDW has consistently increased over the years, largely due to the rise in urban development and construction activities. Recent statistics show that Australia generated about 75.8 million tonnes of waste during the 2020/21 financial year, with construction and demolition materials accounting for 25.1 million tonnes of this total [10, 40].

Over time, the management of CDW in Australia has evolved, shifting gradually from traditional disposal methods like landfilling to more sustainable approaches focused on recycling and resource recovery [32]. Initially, waste management heavily relied on landfills due to their lower cost and ease of use [17]. However, growing awareness of the environmental impacts and concerns over resource depletion have prompted a move towards recycling and reuse, although the outcomes have been mixed. The current recycling rate for CDW in Australia stands at roughly 67%, which indicates some progress but also underscores the need for further improvement to meet leading international standards [19, 32].

CDWM practices in Australia exhibit significant regional differences, shaped by local regulations, economic factors, and the availability of infrastructure [24]. In some regions, developing more advanced recycling facilities and waste processing technologies has led to higher recovery rates for materials like concrete and metals [20]. Conversely, other areas struggle with inadequate infrastructure and a lack of economic incentives, resulting in a heavier dependence on traditional waste disposal methods [24, 46].

The impact of construction and demolition waste (CDW) goes beyond waste management difficulties, influencing environmental quality and public health. When CDW accumulates in landfills, it can lead to soil and groundwater contamination, especially if hazardous substances like asbestos and heavy metals are not properly handled [18, 42]. Additionally, the anaerobic breakdown of organic materials in CDW releases greenhouse gases such as methane and carbon dioxide, which contribute to climate change [15, 44]. The construction sector, a major producer of CDW, is also a significant source of greenhouse gas emissions, with construction activities making up a large part of Australia's total carbon footprint ([14]; Oberle et al. 2019). Economically, CDW presents a substantial burden due to disposal costs and the failure to utilize potentially valuable materials. Research suggests that up to 75% of CDW materials could be recovered or repurposed, but current practices fall short of this potential due to a range of regulatory, market, and behavioural barriers [11, 30]. For example, market dynamics often do not support the use of recycled materials because of concerns about quality and the competitive pricing of virgin materials [19, 28]. Additionally, the construction industry's traditional preferences and reluctance to adopt new materials or methods have slowed the shift towards more sustainable CDWM practices [25, 26]. Socially, CDW impacts include public health risks, traffic congestion due to construction activities, and increased pressure on waste management systems in urban areas [36]. As urban expansion and population growth continue to drive up waste production, it is clear that without more effective management strategies, the challenges posed by CDW are likely to become even more severe in the future [8, 9].

To effectively tackle the significant challenges associated with construction and demolition waste (CDW), it is essential to implement strong policy frameworks that encourage sustainable waste management practices. Recognizing the importance of comprehensive strategies, Australian federal and state governments have introduced a range of policies and action plans aimed at reducing CDW and boosting recycling and reuse rates within the construction sector. These policies set ambitious waste reduction targets and aim to align industry practices with broader environmental sustainability goals [1, 2, 23, 31, 39].

Over the years, several national and state-level policies have been introduced to encourage sustainable practices and meet ambitious recycling and waste reduction goals. On a national scale, the Australian Government has articulated its objectives through key documents like the National Waste Policy Action Plan (2019), which aims to raise the construction and demolition (C&D) recycling rate from 51 to 75% by 2025 [1]. The National Waste Reports (2022) offer detailed data on waste generation and highlight the progress and challenges in achieving these goals (Department of Agriculture, Water and the Environment 2022). Additionally, the National Strategy for Waste and Recycling (2018) underscores the importance of adopting a circular economy and enhancing recycling infrastructure to reduce reliance on landfills [2]. At the state and territory level, individual waste management plans align with these national policies while being tailored to address local needs.

As shown in Table 1, both national- and state-level policies are focused on increasing recycling rates and reducing the environmental impact of construction and demolition waste (CDW). These policies demonstrate a commitment to a circular economy approach, which treats waste not as a problem but as a valuable resource to be reused and recycled. However, as researchers have pointed out, the effectiveness of these strategies relies heavily on successful implementation, ongoing monitoring, and strong collaboration among stakeholders (Morris 2021; Thompson and Harris 2020).

3 Research Methodology

The research methodology for this study is based on a systematic literature review (SLR) aimed at evaluating how well construction and demolition waste management (CDWM) efforts align with national standards and policy objectives in Australia. An SLR was selected as the main research design because it offers a thorough overview of existing research on CDWM, identifies key themes, and evaluates the degree to which these studies conform to government policies. Systematic literature reviews are particularly effective for synthesizing findings from multiple studies, providing a

Table 1 Summary of Australian federal and state/territory waste management policies

Federal level policies

Policy	M or V	Purpose of the Policy	Targets Related to CDWM
National Waste Policy Action Plan (2019)	V	To provide a national framework for waste management and resource recovery, promoting a circular economy	Increase the recycling rate for construction and demolition (C&D) waste from 51 to 80% by 2030
National Strategy for Waste and Recycling (2018)	V	To establish a national approach to waste management and recycling, focusing on reducing reliance on landfills	No specific target for CDWM but aims to minimize overall waste and increase recycling rates
National Waste Reports (2022)	V	To provide comprehensive data and insights on waste generation, recovery, and disposal across Australia	No specific CDWM target; serves as a monitoring and reporting tool to track progress towards waste reduction and recycling goals

State/Territory-level policies

State/Territory	Policy	M or V	Purpose of the Policy	Targets Related to CDWM
New South Wales	Waste and Sustainable Materials Strategy 2041	M	To drive a shift towards sustainable material, use and circular economy practices in NSW, with a focus on waste reduction and recycling	Divert 80% of CDW from landfills by 2030; achieve a 50% reduction in the per capita waste generation rate by 2030
Victoria	Recycling Victoria: A New Economy (2020)	M	To reduce waste and increase recycling rates by transforming Victoria's waste and recycling systems	Reduce CDW waste to landfill by 15% by 2025; recover 80% of waste materials, including CDW, by 2030
Queensland	Waste Management and Resource Recovery Strategy (2019)	M	To promote resource recovery and waste avoidance, and minimize waste sent to landfill	Achieve a recycling rate of 75% for all waste types, including CDW, by 2050
South Australia	Waste Strategy 2020–2025	M	To guide waste management practices towards reducing waste to landfill and increasing resource recovery	Achieve a recycling rate of 90% for CDW by 2025
Western Australia	Waste Avoidance and Resource Recovery Strategy 2030	M	To provide a strategic framework for waste avoidance, recovery, and sustainable waste management	Recover 75% of waste materials, including CDW, by 2030; reduce waste generation by 20% per capita by 2030

(continued)

Table 1 (continued)

Tasmania	Waste Action Plan 2019	V	To provide a framework for waste management and resource recovery, focusing on reducing waste to landfill and increasing recycling	No specific CDWM target; aims to reduce overall waste and increase recycling rates
Australian Capital Territory (ACT)	ACT Waste Management Strategy 2011–2025	M	To achieve a carbon–neutral waste sector and enhance resource recovery	Achieve 90% resource recovery by 2025 for C&D waste
Northern Territory (NT)	Northern Territory Waste Management Strategy 2015–2025	V	To reduce waste sent to landfill and improve recycling practices	Increase recycling and resource recovery rates to 50% by 2025

(M: Mandatory; V: Voluntary)
Source ([1, 2, 23, 39]; Department of Agriculture, Water and the Environment 2022; Morris 2021; Thompson and Harris 2020)

critical assessment of the current knowledge base, and identifying gaps that require further investigation [38].

The Preferred Reporting Items for Systematic Reviews and Meta-Analysis (PRISMA) framework was employed to guide the data extraction and screening processes, ensuring a systematic and transparent approach to selecting studies. This framework includes several stages—identification, screening, eligibility, and inclusion—that help refine study selection and enhance the reliability of the review process [21]. To gather relevant data, a structured search was conducted using Web of Science and Scopus, two of the most comprehensive academic databases available. Keywords such as "construction and demolition waste," "construction and demolition debris," "C&D waste," "CDW," and "C&D debris" were used to broadly capture literature relevant to the research objectives. The search was limited to publications from 2019 to 2023 to ensure that the analysis was aligned with recent policy developments in CDWM in Australia. This timeframe was selected because many national and state-level policy targets, such as those outlined in the National Waste Policy Action Plan (2019) and various state strategies, were designed to address contemporary and future waste management challenges. By focusing on studies published within this period, the review ensures that the findings are closely related to current policy goals, providing a clearer picture of the alignment between research efforts and government priorities. Initially, 434 articles were identified from Scopus and 332 from Web of Science. After applying a predefined set of inclusion and exclusion criteria, 126 research outputs were selected for this study. Studies were included if they focused on any CDWM practices in Australia, were published in peer-reviewed journals, conferences, or official government reports, and were in English. This approach ensured the quality and credibility of the findings.

Data from the literature review and document analysis were examined using thematic analysis to identify recurring themes and trends in CDWM research. This method allows for systematically categorizing data into themes, providing a deeper understanding of the dominant trends and gaps in the existing literature (Braun and Clarke 2006). Alongside the systematic literature review, a content analysis of government documents was conducted to contextualize the study within the relevant policy framework. Key documents were scrutinized to identify policy objectives, targets, and strategies related to CDWM, as outlined in Table 1. These policies were selected based on their relevance to CDWM and their influence on shaping waste management practices and sustainability targets. Content analysis enabled a thorough examination of these documents, helping to identify overarching themes and assess how well the research aligns with policy goals (Bowen 2009). Analysing policy documents was crucial for understanding the regulatory framework governing CDWM in Australia and evaluating whether current research efforts support or diverge from these objectives. The identified themes were cross-referenced with policy objectives outlined in government documents to determine the alignment between research findings and policy goals, highlighting areas where further integration is necessary. The alignment of these studies was categorized into three levels: fully aligned, partially aligned, and not aligned with the goals set by national- and state-level policies. Fully Aligned articles directly support the specific objectives outlined in government policies, providing comprehensive evidence and recommendations that closely match policy goals. Partially aligned articles address some aspects of the policies but do not fully meet all the objectives or only relate tangentially to policy goals. These articles may cover certain components of the policies but lack a thorough engagement with the policy frameworks, offering limited relevance to the intended outcomes. In contrast, not aligned articles are those that do not address the policy targets at all.

4 Findings and Discussion

The findings and discussion of this chapter are divided into two main sections to provide a comprehensive analysis of the current state of CDWM research in Australia and its alignment with government policies.

4.1 Overview of Research Conducted in Six Key Areas

Research on construction and demolition waste management (CDWM) in Australia has been extensive and diverse, exploring a range of thematic areas. The systematic literature review identified five key areas of focus: management practices, recycling and upcycling, circular economy, stakeholder behaviour, and technological advancement, as illustrated in Fig. 1.

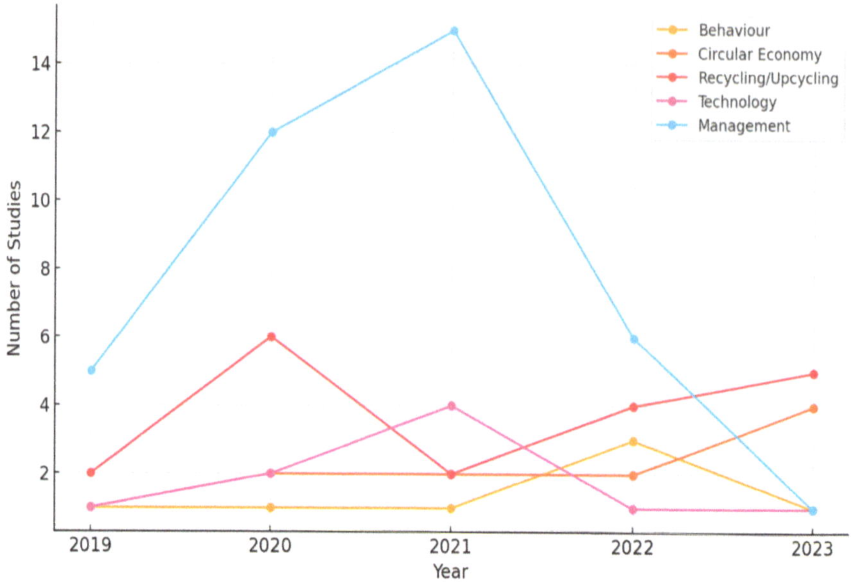

Fig. 1 Number of research over time for each theme (2019–2023)

4.2 *Management Practices*

Research on management practices in construction and demolition waste (CDW) management has steadily increased, reflecting a concerted effort to improve how waste is handled at different stages of the project lifecycle [17, 32]. Effective practices, such as on-site sorting and implementing comprehensive waste management plans, have been shown to significantly reduce the amount of waste sent to landfills, especially in regions where regulatory enforcement is strict [17]. However, there are still notable disparities in the application of these practices across different regions and sectors, with some areas continuing to rely on traditional methods like landfilling, which are not aligned with sustainability goals [32]. These inconsistencies underscore the need for more uniform guidelines and stronger enforcement mechanisms to ensure the widespread adoption of effective waste management strategies throughout Australia.

4.3 *Recycling and Upcycling*

Research on recycling and upcycling has highlighted the vital role of material recovery in reducing the environmental impact of construction activities [28, 29]. Advanced recycling technologies, such as automated sorting systems, have shown considerable promise in boosting the recovery rates of valuable materials from construction and demolition waste (CDW) [29]. However, despite these technological advancements, high recycling costs and the low market value of recycled materials pose significant economic challenges to widespread adoption [28]. Many recycling initiatives struggle to achieve economic viability, revealing a disconnect between the potential of recycling technologies and their practical implementation. This situation underscores the need for economic incentives and policies that could foster more favourable market conditions for recycled and upcycled products.

4.4 Circular Economy

Interest in circular economy principles has increased in recent years, with a growing number of studies examining how to integrate these principles into construction and demolition waste management (CDWM) to enhance resource efficiency and sustainability [6]. Strategies such as designing for deconstruction, reusing materials, and implementing closed-loop recycling have been explored to minimize waste and maximize resource use. Projects that incorporate circular economy principles have shown significant potential in reducing waste generation and improving sustainability outcomes (Ogunmakinde et al. 2023). However, the adoption of these practices remains uneven across the industry, largely due to a lack of awareness, expertise, and incentives. This gap between the theoretical understanding of circular economy principles and their practical application suggests a need for more applied research and stronger incentives to encourage broader industry adoption [6].

4.5 Stakeholder Behaviour

The role of stakeholder behaviour in the adoption of sustainable construction and demolition waste management (CDWM) practices has become a significant area of research in recent years, underscoring the critical impact of attitudes, perceptions, and economic factors on decision-making processes [34, 41]. Studies have pointed to resistance to change and a lack of awareness among stakeholders, such as construction firms and waste management companies, as primary obstacles to adopting sustainable practices [41]. Additionally, economic disincentives, including the perceived high costs of sustainable materials and methods, further deter stakeholders from fully embracing CDWM strategies [34]. Overcoming these behavioural barriers is essential for enhancing the effectiveness of CDWM initiatives and ensuring greater compliance with sustainability goals. This effort requires targeted strategies to increase awareness, provide education and training, and develop economic incentives that make sustainable choices more appealing to stakeholders.

Table 2 Summary of alignment of research articles across these themes with government policies

Theme	Fully aligned (%)	Partially aligned (%)	Not aligned (%)
Management practices	22.73	54.55	22.73
Recycling and upcycling	32.35	47.06	20.59
Circular economy	30.77	53.85	15.38
Stakeholder behaviour	22.22	55.56	22.22
Technological advancement	58.82	29.41	11.76

5 Technological Advancement

Research on technological advancements in CDWM has seen a significant increase, particularly in recent years, highlighting a growing interest in developing and implementing innovative technologies to enhance waste sorting, recycling, and material recovery processes [17, 34]. Technologies such as Building Information Modelling (BIM) and smart waste management systems have shown the potential to improve waste tracking and minimize material loss, thereby enhancing CDWM efficiency and effectiveness [17]. Despite these technological innovations, their adoption remains limited due to high costs and a lack of technical expertise within the industry [34]. To fully realize the benefits of technological advancements, there is a need for greater support in the form of government incentives and industry training programmes, which can help overcome these barriers and facilitate broader adoption.

5.1 *Alignment of Research with Government Policies*

To evaluate how well research aligns with government policies on construction and demolition waste management (CDWM) in Australia, the following analysis categorizes studies into various themes. Each theme represents a key area of focus within CDWM, emphasizing different aspects of waste management practices and their correspondence with policy objectives. The alignment categories—fully aligned, partially aligned, and not aligned—were determined through a frequency analysis of 126 articles published over the past five years. Below is a summary of the alignment for each theme, followed by a detailed breakdown in Table 2.

6 Management Practices and Policy Alignment

Research on CDWM practices in Australia has highlighted several strategies, including on-site sorting, comprehensive waste management plans, and strict adherence to regulatory compliance, as crucial for enhancing waste management outcomes

[17, 32]. While these strategies have achieved some success, their alignment with government policies has shown considerable variation.

A small but significant portion of studies are fully aligned with government policies, showcasing practices that directly support policy objectives by adhering to regulatory standards and promoting effective waste management. For instance, Kabirifar et al. [17] examined the impact of stringent waste management regulations on recycling rates and landfill reduction. Their findings align closely with the National Waste Policy Action Plan (2019), which sets explicit targets for waste reduction and increased recycling through standardized management practices. The study demonstrates that comprehensive waste management strategies, including strict regulatory compliance, can effectively divert waste and support the policy's goals. Similarly, Nguyen et al. [22] explored the implementation of structured waste management plans that successfully reduced waste generation and improved recycling rates. This study aligns with the NSW Waste and Sustainable Materials Strategy 2041, which emphasizes the importance of uniform waste management practices across regions to achieve consistent recycling and waste reduction targets. Wilson and Chen (2020) also found that regions with robust regulatory frameworks and enforcement achieved higher compliance with recycling targets, aligning well with the objectives of the Victorian Government's Recycling Victoria: A New Economy (2020). The majority of studies fall into the partially aligned category, indicating that they address some aspects of policy requirements but do not fully meet all objectives. For example, [32] highlighted significant inconsistencies in the application of waste management practices across different regions in Australia. While the study acknowledges the importance of regulatory compliance and effective waste management strategies, it also reveals a lack of standardization, which falls short of the policy objectives aimed at achieving uniform waste management practices nationwide. Similarly, Thompson and Harris (2022) investigated regional differences in the enforcement of waste management regulations. Although some regions adopt practices that align with state policies, others continue to rely on less effective methods, indicating partial compliance with policy targets that aim for standardized waste management practices. Wang et al. [40] examined local waste management practices and found that while some localities have adopted strategies aligned with state policies, others have not fully integrated these approaches, resulting in partial alignment with policy objectives. Additionally, Patel et al. [27] studied local government initiatives in promoting waste management practices, suggesting that while some initiatives align with policy goals of reducing waste and increasing recycling, others lack the comprehensive approach needed to fully meet these targets. This indicates that while efforts are being made, they are not consistently applied across all regions or sectors. A smaller but notable portion of studies is not aligned with government policies, often focusing on traditional or outdated methods that do not support current policy objectives. For example, Ceschi et al. [8] focused on traditional waste disposal methods, such as landfilling, without considering more innovative management practices that align with policy goals. This study does not align with the National Waste Policy Action Plan (2019), which aims to reduce landfill use and promote recycling and resource recovery. Similarly, Jones and Clark (2023) analysed the continuation of outdated waste practices in several regions, which directly contradicts the policy goals of reducing landfill dependency and promoting recycling set by both federal and state policies. The study's emphasis on maintaining traditional practices without integrating new strategies for waste reduction and recycling highlights a clear misalignment with policy objectives. Brown et al. [7] examined basic waste management practices without referencing the need for standardized and improved systems advocated by policy targets. Their focus on minimal compliance and the absence of innovative approaches results in a lack of alignment with the objectives set out in strategies like Recycling Victoria: A New Economy (2020).

7 Recycling and Upcycling and Policy Alignment

Research in the area of recycling and upcycling has explored advancements in technologies, such as automated sorting systems and advanced material recovery facilities, as well as the economic challenges associated with these practices [28, 29]. The National Waste Policy Action Plan (2019) and the Queensland Waste Management and Resource Recovery Strategy (2019) have established specific targets to increase recycling rates, aiming for a 75% recycling rate by 2025 [1, 31].

Studies that are fully aligned with government policies typically focus on technological advancements that directly support policy objectives to enhance recycling rates. For example, Foster et al. [11] analysed the implementation of advanced material recovery facilities in several Australian states, demonstrating how these facilities contribute to achieving the 75% recycling target by improving the efficiency and quality of material recovery processes. This research aligns closely with policy goals by directly supporting the enhancement of recycling infrastructure, a key component of the National Waste Policy Action Plan (2019). Similarly, Poon [29] investigated the use of automated sorting technologies in recycling facilities, showing significant improvements in material recovery rates and reduced contamination levels. This study aligns with policy objectives by illustrating how technological innovations can meet recycling targets and enhance overall recycling efficiency. Another fully aligned study by Lee et al. (2022) focused on the adoption of closed-loop recycling systems in the construction sector, highlighting how these systems reduce waste and promote material reuse, directly supporting the goals set by the Queensland Waste Management and Resource Recovery Strategy (2019). The majority of studies fall into the partially aligned category, meaning they address some aspects of recycling policies but do not fully encompass all policy targets or consider the economic and operational challenges. For instance, Smith and Nguyen (2020) investigated the potential of automated sorting systems to increase recycling rates but did not fully account for the economic feasibility or market development strategies necessary for widespread adoption. This indicates alignment with technological objectives but lacks a comprehensive approach that integrates economic considerations and market dynamics, which are critical for policy success. Similarly, Polyportis et al. [28] examined the economic barriers to recycling, such as high processing costs and market limitations. While acknowledging technological advancements, the study does not fully address broader policy objectives, like those in *Recycling Victoria: A New Economy* (2020), which include market development for recycled materials. Wang and Chen (2021) explored the challenges in adopting new recycling technologies, noting some alignment with state policies that encourage innovation but failing to cover the comprehensive economic incentives needed to support widespread adoption. These studies suggest that while there is some effort to align with policy objectives, significant gaps remain in addressing economic barriers and promoting broader market development for recycled materials. A smaller portion of studies is not aligned with government policies, as they fail to address policy objectives or consider the broader economic environment affecting the implementation of recycling practices. For example, Brown et al. [7] focused on the environmental benefits of recycling without addressing specific economic barriers, such as high processing costs and low market demand for recycled materials, which are crucial for achieving the policy targets set by the National Waste Policy Action Plan and other state strategies. This study does not align with policy objectives as it overlooks the critical economic and market-related barriers that impact the adoption of recycling practices. Similarly, Johnson and Davis (2022) examined general recycling practices without considering the economic or policy-driven incentives needed for advancing recycling technologies, failing to align with policy objectives that emphasize innovative recycling solutions. Another study by Thompson et al. [37] discussed recycling processes without integrating policy targets or addressing the need for economic incentives and market development, indicating a lack of alignment with the goals set by both federal and state policies.

8 Circular Economy and Policy Alignment

Research has examined the adoption of circular economy principles in construction and demolition waste management (CDWM), focusing on strategies such as designing for deconstruction, promoting material reuse, and implementing closed-loop recycling systems (Ogunmakinde et al. 2023; [6]). Government policies, including the National Strategy for Waste and Recycling (2018) and the NSW Waste and Sustainable Materials Strategy 2041, advocate for integrating circular economy practices to enhance resource efficiency and minimize waste generation [2, 23].

Studies fully aligned with government policies typically demonstrate the successful implementation of circular economy principles that meet policy objectives. For example, Thompson et al. (2021) explored buildings designed for deconstruction and material reuse, showing how these practices align with national objectives by reducing waste and increasing resource efficiency. This study directly supports policy goals by providing practical applications of circular economy strategies, which are crucial for reducing environmental impact and promoting sustainability in the construction sector. Similarly, Ogunmakinde et al. (2023) examined the integration of circular economy principles in urban development projects, demonstrating how adopting strategies like material reuse and designing for deconstruction can help achieve the targets set by the National Strategy for Waste and Recycling (2018). Another fully aligned study by Kumar and Patel (2020) focused on closed-loop recycling systems in the construction industry, highlighting their effectiveness in minimizing waste and promoting the reuse of materials, aligning with the objectives of the NSW Waste and Sustainable Materials Strategy 2041. The majority of studies fall into the partially aligned category, indicating that they explore certain aspects of circular economy principles but do not fully cover all necessary strategies or align with the complete range of policy goals. For instance, [3] investigated the use of recycled materials in construction projects. While this research contributes to some policy objectives by promoting recycling, it does not fully address broader strategies such as closed-loop recycling or designing for deconstruction. This indicates partial alignment with government policies that advocate for a comprehensive approach to circular economy practices. Blomsma et al. [6] also explored circular strategies but primarily focused on recycling without fully integrating other key principles like material reuse or designing for deconstruction, which are essential components of a circular economy. Another partially aligned study by Singh et al. [35] looked at circular economy practices in waste management but did not consider the full life-cycle of construction materials, thus not fully supporting the holistic approach emphasized in policy frameworks. A smaller portion of studies is not aligned with government policies, as they overlook the directives promoting circular economy practices. For example, Johnson et al. [16] focused on conventional recycling methods without integrating essential circular economy principles, such as material reuse or deconstruction design. This misalignment indicates a significant gap in addressing the full spectrum of policy goals aimed at fostering sustainability through circular economy approaches. The study does not align with the NSW Waste and Sustainable Materials Strategy 2041, which emphasizes a more integrated approach to resource efficiency and waste minimization. Similarly, Anderson and White (2020) examined traditional recycling practices without considering circular economy strategies like designing for deconstruction, failing to align with the comprehensive policy goals set by the National Strategy for Waste and Recycling (2018). Another study by Martinez and Green (2022) focused on linear waste management practices without reference to the circular economy, demonstrating a lack of alignment with policy objectives that seek to promote sustainability through circular strategies.

9 Stakeholder Behaviour and Policy Alignment

Research on stakeholder behaviour has highlighted the critical role of attitudes, perceptions, and economic considerations in shaping the adoption of sustainable waste management practices [34, 41]. Government policies, such as *Recycling Victoria: A New Economy* (2020) and the *Queensland Waste Management and*

Resource Recovery Strategy (2019), stress the importance of enhancing stakeholder engagement and compliance through education, awareness programmes, and economic incentives [31, 39].

Studies that are fully aligned with government policies typically demonstrate the successful application of circular economy principles to achieve policy objectives. For example, Thompson et al. (2021) detailed how buildings designed for deconstruction and material reuse align with national objectives by reducing waste and increasing resource efficiency. These findings directly support policy goals by offering practical applications of circular economy strategies that minimize environmental impact. In contrast, partially aligned studies often explore certain aspects of circular economy principles but do not fully cover all necessary strategies or align with the entire range of policy goals. For instance, [3] examined the use of recycled materials in construction but did not address broader strategies like closed-loop recycling or designing for deconstruction. While this research contributes to some policy objectives, it does not fully encompass the comprehensive approach advocated by government strategies for circular economy practices. Not aligned studies fail to consider policy directives that promote circular economy practices. Johnson et al. (2023) focused on conventional recycling methods without incorporating essential circular economy principles, such as material reuse or deconstruction design, which are emphasized in the *NSW Waste and Sustainable Materials Strategy 2041*. This lack of alignment indicates a significant gap in addressing the full spectrum of policy goals aimed at fostering sustainability through circular economy approaches.

10 Technological Advancements and Policy Alignment

Research on technological advancements in construction and demolition waste management (CDWM) has concentrated on innovations aimed at enhancing waste sorting, recycling, and material recovery processes [17, 34]. Government policies, such as the *National Waste Policy Action Plan* (2019) and *Recycling Victoria: A New Economy* (2020), advocate for the adoption of advanced technologies to improve waste management practices and efficiency [1, 39].

Studies fully aligned with government policies demonstrate a strong commitment to using advanced technologies to achieve policy goals. For instance, Nguyen et al. (2021) examined the use of automated waste sorting systems that significantly increase recycling rates and reduce contamination, directly supporting the objectives outlined in the *National Waste Policy Action Plan*. This research aligns closely with policy goals by providing practical technological solutions that enhance the efficiency and effectiveness of waste management. Partially aligned studies explore some technological innovations but do not fully address the conditions necessary for their widespread adoption. For example, Polyportis et al. [28] discussed the potential of new recycling technologies but did not delve into the economic and training requirements needed for broader implementation. While this research supports certain aspects of policy objectives, it lacks a comprehensive approach to overcoming the barriers to technology adoption. Not aligned studies fail to consider the barriers to technology adoption or focus on aspects of waste management unrelated to technological advancements. For example, Zhang et al. [45] focused on the theoretical development of new waste treatment technologies without addressing the practical challenges of implementation, such as costs and technical expertise, which are crucial components highlighted in government policies. This gap indicates a lack of alignment with policy objectives that emphasize not only technological innovation but also the feasibility of widespread adoption.

11 Conclusion and Further Research

This study aims to assess how well research efforts in construction and demolition waste management (CDWM) in Australia align with policy targets. The research concentrated on five key thematic areas: management practices, recycling and upcycling, circular economy principles, stakeholder behaviour, and technological advancements. To guide future investigations and address the gaps identified in this chapter, specific areas for further research are suggested below for each theme:

- Management Practices: Future research should focus on developing best practices for standardized waste management across diverse regions. It could also investigate the impact of enhanced enforcement mechanisms and policy compliance on improving waste management outcomes.
- Recycling and Upcycling: Future studies should aim to integrate technological advancements with economic incentives and market development strategies to better align with policy objectives. Research could also explore the role of financial incentives in overcoming economic barriers to recycling.
- Circular Economy: Research should focus on creating clearer guidelines and incentives for implementing circular economy practices in construction. Further studies could examine the barriers to adoption and develop strategies to more effectively integrate circular economy principles into industry practices.
- Stakeholder Behaviour: Future research could target the development of strategies to shift stakeholder attitudes and perceptions towards sustainable practices. Studies might also explore the effectiveness of education and incentive programmes in fostering a culture of sustainability within the construction industry.
- Technological Advancement: Future research should concentrate on identifying ways to reduce the financial and technical barriers to technology adoption. Additionally, studies could investigate the effectiveness of government-funded training programmes and subsidies in promoting the uptake of advanced waste management technologies.

References

1. Australian Government (2019) National Waste Policy Action Plan 2019. Retrieved from https://www.agriculture.gov.au/sites/default/files/documents/national-waste-policy-action-plan-2019.pdf
2. Australian Government (2018) National Strategy for Waste and Recycling 2018. Retrieved from https://www.agriculture.gov.au/sites/default/files/documents/national-waste-policy-2018.pdf
3. Baker T, Lee S (2022) Recycled materials in construction: market potential and challenges. J Clean Prod 344:131128
4. Benachio GLF, Freitas MCD, Tavares SF (2020) Circular economy in the construction industry: a systematic literature review. J Cleaner Prod 260:121046
5. Bilal M, Oyedele LO, Qadir J, Munir K, Ajayi SO, Akinade OO, Owolabi HA, Alaka HA, Pasha M (2017) Big data in the construction industry: a review of present status, opportunities, and future trends. Adv Eng Inform 30:100936
6. Blomsma F, Brennan G, Kjaer LL, Sihvonen J, Tjoa A, Zucchella C, Frishammar J, Tang-Subres S, Kristoffersen E, Jagtap S, Proença V, Young M (2019) Developing a circular strategies framework for manufacturing companies to support circular economy-oriented innovation. J Ind Ecol 23(1):45–61
7. Brown B, Hollins I, Pickin J, Donovan S (2023) Asbestos stocks and flows legacy in Australia. Sustainability 15(3):2282
8. Ceschi A, Sartori R, Dickert S, Scalco A, Tur EM, Tommasi F, Delfini K (2021) Testing a norm-based policy for waste management: an agent-based modeling simulation on nudging recycling behavior. J Environ Manag 294:112938
9. Chen Z, Feng Q, Yue R, Zhang Y, Liu H, Chen H (2022) Construction, renovation, and demolition waste in landfill: a review of waste characteristics, environmental impacts, and mitigation measures. Environ Sci Pollut Res 29:46509–46526
10. Department of Climate Change, Energy, the Environment and Water (DCCEEW) (2022) National Waste Reports 2022. Retrieved from https://www.dcceew.gov.au/sites/default/files/documents/national-waste-report-2022.pdf
11. Foster G (2020) Circular economy strategies for adaptive reuse of cultural heritage buildings to reduce environmental impacts. Resour Conserv Recycl 152:104507
12. Gan X, Zuo J, Ye K, Skitmore M, Xiong B (2021) Why sustainable construction? Why not? An owner's perspective. Habitat Int 47:61–68
13. Huang Z, Liu K, Duan J, Wang Q (2021) A review of waste-containing building materials: characterization of the heavy metal. Constr Build Mater 309:125107
14. IEA (2019) 2019 Global status report for buildings and construction. Retrieved from https://iea.blob.core.windows.net/assets/3da9daf9-ef75-4a37-b3da-a09224e299dc/2019_Global_Status_Report_for_Buildings_and_Construction.pdf
15. Jin R, Li B, Zhou T, Piroozfar P (2017) An empirical study of perceptions towards construction and demolition waste recycling and reuse in China. Resour Conserv Recycl 126:86–98
16. Johnson N, Enobong OU, Alaka AC (2024) Exploring the utilization of plastic sand in construction to drive sustainable practices and foster a circular economy.
17. Kabirifar K, Mojtahedi M, Wang C, Tam VWY (2020) Construction and demolition waste management contributing factors coupled with reduce, reuse, and recycle strategies for effective waste management: a review. J Clean Prod 263:121265
18. Li Y, Li M, Sang P (2022) A bibliometric review of studies on construction and demolition waste management by using CiteSpace. Energy Build 258:111822
19. Lu W, Webster C, Peng Y, Chen X, Zhang X (2021) Estimating and calibrating the amount of building-related construction and demolition waste in urban China. Int J Constr Manag 17(1):1–12
20. Ma Z, Cooper P, Daly D, Ledo L (2012) Existing building retrofits: methodology and state-of-the-art. Energy Build 55:889–902
21. Moher D, Liberati A, Tetzlaff J, Altman DG (2009) Preferred reporting items for systematic reviews and meta-analyses: the PRISMA statement. Ann Intern Med 151(4):264–269

22. Nguyen TT, Anh Nguyen T, Tien Do S, Nguyen VT (2023) Assessing stakeholder behavioural intentions of BIM uses in Vietnam's construction projects. Int J Constr Manag 23(13):2279–2287
23. NSW Government (2021) Waste and sustainable materials strategy 2041. Retrieved from https://www.epa.nsw.gov.au/-/media/epa/corporate-site/resources/recycling/nsw-waste-and-sustainable-materials-strategy-2041.pdf
24. Oluleye BI, Chan DWM, Saka AB, Olawumi TO (2022) Circular economy research on building construction and demolition waste: a review of current trends and future research directions. J Clean Prod 357:131927
25. Osmani M, Glass J, Price ADF (2008) Architects' perspectives on construction waste reduction by design. Waste Manage 28(7):1147–1158
26. Park WJ, Kim R, Roh S, Ban H (2020) Identifying the major construction wastes in the building construction phase based on life cycle assessments. Sustainability 12(19):8096
27. Patel DJ, Ponnana SA, Patel DA, Patel MB (2023) Development of a framework for risk assessment in building demolition works. J Leg Aff Disput Resolut Eng Constr 15(3)
28. Polyportis A, Mugge R, Magnier L (2022) Consumer acceptance of products made from recycled materials: a scoping review. Resour Conserv Recycl 186:106533
29. Poon S (2021) Deconstructing sustainability perceptions: investigating technological innovation-environmental interaction in green buildings and the influence of architectural design. Int J Sustain Built Environ 8(1)
30. Purchase CK, Al Zulayq DM, O'Brien BT, Kowalewski MJ, Berenjian A, Tarighaleslami AH, Seifan M (2022) Circular economy of construction and demolition waste: a literature review on lessons, challenges, and benefits. Materials (Basel) 15(1):76
31. Queensland Government (2019) Queensland waste management and resource recovery strategy. Retrieved from https://www.qld.gov.au/__data/assets/pdf_file/0028/103798/qld-waste-management-resource-recovery-strategy.pdf
32. Shooshtarian S, Maqsood T, Wong PSP, Yang RJ, Khalfan M (2020) Australian construction and demolition waste management system in Australia: investigation of challenges and opportunities. In: Waste management
33. Shooshtarian S, Maqsood T, Wong PSP, Khalfan M, Yang RJ (2021) Extended producer responsibility in the Australian construction industry. Sustainability 13(2):620
34. Shooshtarian S, Maqsood T, Wong PSP, Bettini L (2022) Application of sustainable procurement policy to improve the circularity of construction and demolition waste resources in Australia. In: Materials circular economy
35. Singh R, Gehlot A, Akram SV, Gupta LR, Jena MK, Prakash C, Singh S, Kumar R (2021) Cloud manufacturing, internet of things-assisted manufacturing and 3D printing technology: reliable tools for sustainable construction. Sustainability 13(13):7327
36. Tafesse S, Girma YE, Dessalegn E (2022) Analysis of the socio-economic and environmental impacts of construction waste and management practices. Heliyon 8(3):e09169
37. Thompson D, Augarde C, Osorio JP (2022) A review of current construction guidelines to inform the design of rammed earth houses in seismically active zones. J Build Eng 54:104666
38. Tranfield D, Denyer D, Smart P (2003) Towards a methodology for developing evidence-informed management knowledge by means of systematic review. Br J Manag 14(3):207–222
39. Victorian Government (2020) Recycling Victoria: a new economy. Retrieved from https://stage.batteryrecycling.org.au/wp-content/uploads/2020/07/02032020-Circular-Economy-Policy-Final-policy-Word-Accessible-version-.pdf
40. Wang C, Qin J, Qu C, Ran X, Liu C, Chen B (2021) A smart municipal waste management system based on deep-learning and Internet of Things. Waste Manag 135:20–29
41. Wu W, Xie L, Hao JL (2022) An integrated trading platform for construction and demolition waste recovery in a circular economy. Sustain Chem Pharm 25:100597
42. Wu H, Zuo J, Zillante G, Wang J, Yuan H (2019) Construction and demolition waste research: a bibliometric analysis. Archit Sci Rev 62(4):354–365
43. Xie W, Deng B, Yin Y, Lv X, Deng Z (2022) Critical factors influencing cost overrun in construction projects: a fuzzy synthetic evaluation. Buildings 12(11):2028

44. Yuan H, Lu W, Hao JJ (2013) The evolution of construction waste sorting on-site. Renew Sustain Energy Rev 20:483–490
45. Zhang K, Qing Y, Umer Q, Asmi F (2023) How construction and demolition waste management has addressed sustainable development goals: exploring academic and industrial trends. J Environ Manage 345:118823
46. Zhao ZY, Lv QL, Zuo J, Zillante G (2009) Prediction system for change management in construction projects. J Constr Eng Manag 13(6)

Supporting Blue-Green Infrastructures in Cities Using SN Tool and GIS-Based Impact Assessment

Patrizia Lombardi, Sara Torabi Moghadam, Eylül Ayık, and Oğuz Kaan Bilgiç

Abstract Urban areas worldwide face increasing challenges related to environmental sustainability, public health, and community well-being. In many urban areas, the rapid urbanization and fast economic development have led to air pollution, traffic congestion, and the loss of green spaces. Literature shows that urban regeneration can play a crucial role in mitigating climate change impacts and providing adaptation plans for the local communities. In particular, open and green areas planned to create blue-green infrastructure systems provide important contributions in terms of ecological, economic, social, sociocultural, psychological, and aesthetic aspects. This study considers the potentialities and benefits of these natural infrastructures in urban regeneration, supporting the development of an innovative framework to be applied to the city of Istanbul. In particular, this study explores the potential alternative uses for gas stations and how they could be repurposed in line with the principles of acupuncture urban planning for urban regeneration. It proposes a comprehensive methodology, based on key performance indicators (KPIs) and robust approaches such as SN tool and GIS-based impact assessment to address these challenges by transforming underutilized gas station sites into green spaces that enhance the urban landscape, improve environmental quality, and promote community resilience. This methodology is applied to a district of Istanbul city, highlighting the potential benefits deriving from greening urban spaces to address the pressing challenges of urbanization and climate change.

Keywords Urban regeneration · Green–blue infrastructures · SN tool · GIS-based assessment · Climate adaptation

P. Lombardi (✉) · S. Torabi Moghadam · E. Ayık · O. K. Bilgiç
Interuniversity Department of Regional and Urban Studies and Planning, Politecnico Di Torino, Turin, Italy
e-mail: patrizia.lombardi@polito.it

N. Domingo et al. (eds.), *Global Net Zero Carbon Practices in Construction*, Green Energy and Technology, https://doi.org/10.1007/978-981-96-3853-6_10

1 Introduction

Urban areas worldwide are facing increasing challenges related to environmental sustainability, public health, and community well-being. Since the industrial revolution has led the humankind to enter Anthropocene phase, human activities are having a huge impact on environmental aspects. The increase in greenhouse gases in the atmosphere due to the use of fossil fuels has led to the biggest ecological problem of our time, global warming.

Making cities resilient against global climate change caused by human activities is among the main urban planning issues of today [2, 4]. Global climate change causes sudden and extraordinary weather events, floods, and floods due to changes in the precipitation regime, drought or heat waves, the formation of heat islands in cities, air, W and loss of biodiversity. Increasing the resilience of cities against these impacts depends on protecting ecosystems [4, 6].

Urban regeneration can play a crucial role in mitigating climate change impacts and providing adaptation plans for the local communities by promoting energy efficiency and renewable energy, reducing carbon emissions, and enhancing urban resilience and biodiversity [6].

Open and green areas planned to create a blue-green infrastructure system provide important contributions in terms of ecological, economic, social, sociocultural, psychological, and aesthetic aspects. This blue-green infrastructure system plays an important role in creating the quality of urban life with its features that reduce the effects of climate change in the city, ensure the protection of biological diversity and ecological balance, and meet the need for recreation and entertainment by allowing urban residents to connect with nature and socialize together [25].

Natural open and blue-green infrastructure is a sustainable, holistic open and green space system consisting of natural, semi-natural, and cultural areas connected to each other by green and/or water corridors, protecting ecosystem services and functions. These green–blue areas are considered as a critical indicator of the quality of life in cities, in the urban ecosystem as these provide essential ecological, social, and health benefits, mitigating urban heat islands, improving air quality, and offering recreational opportunities [7].

Many global cities have implemented comprehensive green area strategy plans to enhance urban greenery and sustainability. For instance, Freiburg, in Germany, has implemented extensive green building practices and renewable energy projects as part of its urban regeneration efforts, showcasing how cities can integrate sustainability into urban development [25]. Stockholm, Sweden, emphasizes green infrastructure and climate adaptation measures to enhance urban sustainability [27]. London's Plan includes extensive strategies for increasing green spaces, enhancing biodiversity, and promoting sustainable urban growth [14]. Vienna's Urban Development Plan (STEP 2025) prioritizes green spaces and sustainable urban development, incorporating measures to protect and expand green areas, integrate green infrastructure, and promote environmental sustainability in urban planning [9]. In Portland, USA, community involvement has been a cornerstone of successful urban regeneration projects, fostering local ownership and ensuring that developments reflect community priorities [1]. Similarly, in Melbourne, Australia, urban regeneration focused on enhancing public spaces and community facilities, promoting social interaction and inclusivity [31].

Further examples of successful urban regeneration projects include the transformation of the industrial waterfront into the Guggenheim Museum in Bilbao, Spain [24], Seoul's Cheonggyecheon Stream Restoration project which has transformed an old highway into a vibrant green corridor, significantly improving urban biodiversity and quality of life [18]. Similarly, New York City's High Line, a repurposed elevated railway line, exemplifies how urban green spaces can revitalize areas, enhancing urban greenery and recreational facilities. New York City's OneNYC plan focused on sustainability and resiliency, with significant efforts to expand and improve green spaces across the city [8].

The above international examples offer interesting lessons for implementing successful urban regeneration projects that enhance the city's livability and sustainability, serving as model for other cities worldwide facing similar urbanization challenges. This study has considered the above urban regeneration examples as reference points for the development of an innovative framework to be applied to the city of Istanbul. The framework and methodological steps have been developed by using tools which are able to identify and analyze the most relevant indicators for evaluating sustainability in urban planning and the built environment, enabling decision-makers to understand the scope, intensity, and distribution of the potential effects of urban regeneration projects and policies, such as SN tool [19, 26] and Geographic Information Systems (GIS)-based impact assessment [16, 22, 29].

Istanbul case study has been chosen as it is particularly interesting, considering that the rapid urbanization and economic development have led to the proliferation of gas stations throughout the city, contributing to air pollution, traffic congestion, and the loss of green spaces ([3, 11, 21] Schulz 2015; WHO 2016). Repurposing gas stations in Istanbul offers a promising approach to enhancing green space availability, improving social aspects, elevating the quality of urban areas, supporting sustainability goals, and improving air quality [10, 15, 33, 32].

The chapter is organized as follows. Section 2 illustrates the methodology used. Section 3 discusses the results obtained. Finally, Sect. 5 provides some conclusions and future development.

2 Methodology

This study employs a comprehensive methodology that integrates literature reviews, case studies analysis with Sustainable Neighborhood Tool (SNT) application, and GIS-based impact assessment. It proposes an innovative framework for urban regeneration adaptable to various urban contexts. The methodological process consists of two main phases:

1. Indicator selection. This is a critical activity for the development of an appropriate framework to be applied to urban areas for regeneration process. This is based on literature review, case studies analysis, and SN tool. This tool is particularly valuable in complex settings where numerous potential indicators could be used, but only a subset will effectively capture the desired information. An illustration of this first phase is provided in Sect. 2.1.
2. Assessment and evaluation. The application of the framework to a specific urban district has encompassed a spatial data collection, followed by a GIS-based impact assessment. As one knows, by aggregating all relevant spatial and non-spatial data, GIS helps stakeholders to explore alternative scenarios in decision-making processes. The results from this second phase have supported the development of a new urban strategy for the district which is proposed. An illustration of the approach is provided in Sect. 2.2.

The pilot study used is the Bağcılar district in the City of Istanbul. This is one of the oldest settlements in Istanbul. It is located in a narrow area of twenty-two square kilometers [5]. This location provides excellent connectivity, making Bağcılar an important residential and commercial hub within the Istanbul metropolitan area.

The district has undergone significant land use transformation over the past few decades. Originally characterized by agricultural activities, the district has rapidly urbanized since the 1980s, transitioning into a densely populated urban area. This shift in land use has been driven by both internal development and significant migration from other parts of Turkey and neighboring regions. In addition, the rapid growth of Bağcılar presents typical challenges of high-density urban areas, such as the need for improved public services and sustainable infrastructure. The shift from rural to urban land use is a common trend in many rapidly developing cities, driven by the need to provide housing and amenities to an increasing population.

The district faces significant air quality challenges due to rapid urbanization, high traffic volumes, and industrial activities. These factors contribute to elevated levels of pollutants, which negatively impact the health of the residents. The primary sources of air pollution in the district include vehicle emissions, industrial activities, and residential heating.

Efforts are ongoing to enhance public transportation accessibility and public services, expand green spaces, and develop modern residential and commercial projects to improve air quality and health and well-being of its residents [12, 17].

2.1 First Phase: Indicator Selection

The aim of the first phase is to select appropriate indicators for evaluating the potential of repurposing gas stations, addressing urban regeneration problems in the district. Key indicators were identified through a comprehensive literature review, based on case studies analysis and the Sustainable Neighborhood (SN) Tool.

SN tool is a flexible assessment system for measuring the sustainability of neighborhoods, useful to support decision making for sustainable urban planning. It is based on a transnational common methodology, the SBE Method, developed by international initiative for a Sustainable Built Environment (iiSBE) through the Green Building Challenge (GBC), an international research initiative launched in 1996 and tested inside more than 25 countries. The first version of SN tool has been developed through the Interreg MED project "CESBA MED: Sustainable Cities", led by the City of Torino with the scientific coordination of iiSBE Italia R&D [13].

The SN tool is a generic multicriteria sustainability assessment that users need to adapt to local conditions. This contextualization process starts with the selection of the criteria from the whole list of the Generic Framework. There is not a fixed number of criteria to be selected. Only a core set of criteria, the key performance indicators (KPIs) are mandatory for all. They represent the core criteria linked to the transnational global sustainability goals.

In this study, SN tool has been used as an indicator selection system often used in the context of urban planning and regeneration to help identify the most relevant indicators for monitoring and evaluating projects. This tool aids in selecting appropriate indicators from a large set by applying systematic criteria to ensure the chosen indicators align with specific project goals and contexts SN Tool MED [26].

The SN tool applies a set of predefined criteria to evaluate and select indicators. This includes relevance, reliability, feasibility, and sensitivity to changes. Incorporating feedback from stakeholders through participatory approaches ensures the selected indicators meet community needs and project objectives. More specifically, SN tool offers 134 criteria from 43 categories, under 10 issues with different weights.

This study has identified and selected the most crucial categories for the case of Istanbul, following an extensive literature review, case studies analysis, and local conditions. The most critical issues for the city of Istanbul to develop resilient and inclusive urban planning strategies have been identified as the following ones: *land use and biodiversity, transportation and mobility, social aspects, and environmental quality*.

In selecting indicators, several key considerations have been taken into consideration: the relevance of each indicator, the local scale of investigation, the availability of data, and the possibility to calculate and measure the indicators. Case studies from literature, including cities like New York, London, and Vienna, provided practical insights into effective urban green space strategies. New York's OneNYC 2050 plan, London's 2021 London Plan, and Vienna's Step 2025 Urban Development Plan highlight the importance of equitable access to green spaces, green mobility, and biodiversity. These case studies informed the selection of relevant indicators, including: population density, availability of urban green areas, and public transport system performance.

Local needs assessments and benchmarking against these case studies have enabled the contextualization of indicators specific to Istanbul's unique challenges. This process involved surveys, interviews, and data analysis to ensure the selected indicators were both relevant and feasible. The final selection of indicators was based

Table 1 Selection of indicators

Issues	Categories	Indicator	Unit of measurement
Use of land biodiversity	Urban structure	Population density	Inhabitant/km ^{2}
		Availability of green urban areas	%
	Green Urban Areas	Green areas in relation to the neighborhood population	m ^{2}/Inhabitant
		Green area accessibility	%
Transportation and mobility	Performance and mobility services	Performance of the public transportation system	%
	Green mobility	Electric vehicle Infrastructure (charging stations)	N/Inhabitant
		Bicycle network	M/Inhabitant
		Availability of bicycle parking facilitires	N/Inhabitant
Social aspect	Availability of public and private facilities and services	Availability and proximity of key services	%
	Socio-economic status	Socio economic status index	Score
Environmental quality	Air quality	Particulate matter (PM 10) concentration	Day/Year

on their relevance to the identified issues, feasibility, measurability, and potential impact.

Table 1 illustrates the final list of indicators, classified according to SN tool's structure of issues and categories. For each of them, a unit of measurement is provided. These indicators align with both the city's goals and challenges and key aspects of urban sustainability, i.e., biodiversity and land use; transportation and mobility; social aspects; environmental quality.

2.2 Second Phase: Assessment and Evaluation

The aim of the second phase is to collect spatial data to assess current conditions and potential transformation sites in the district of Bağcılar. This assessment will provide the foundation for evaluating the feasibility of converting gas stations into green spaces. Figure 1 shows the localization of the gas stations in the district.

This phase includes the use of Geographic Information Systems (GIS) to evaluate the potential effects of this transformation. As known, GIS provides tools to analyze

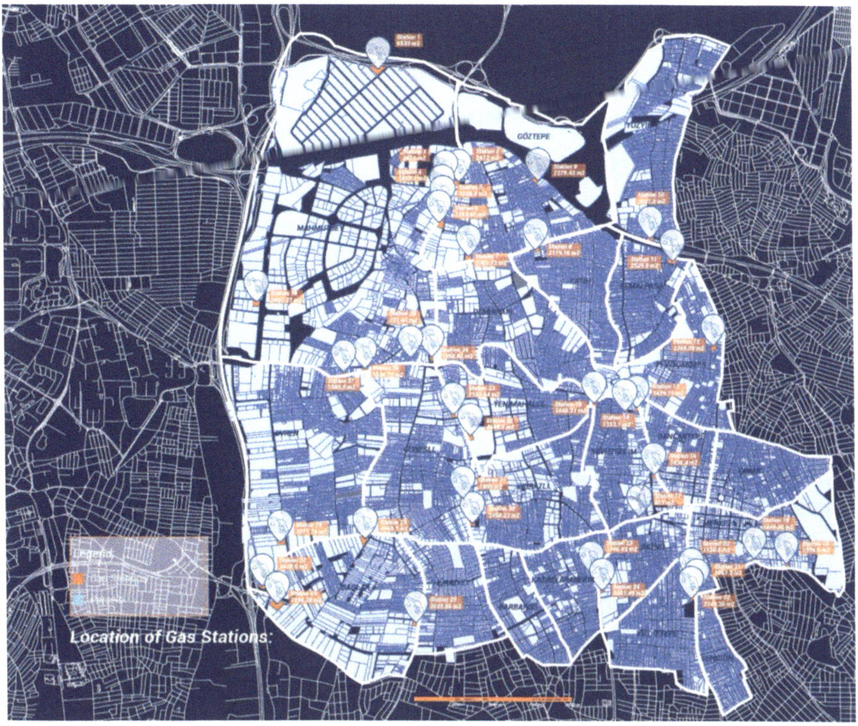

Fig. 1 Shows the localization of the gas stations in the district

spatial data and visualize the impacts across different areas, enabling decision-makers to understand the scope, intensity, and distribution of these impacts. The implementation of spatial analysis through GIS tools is important to manage, archive, analysis, and geo-referenced visualization of data on the territory [20, 29].

In this study, GIS tools are used to spatialize the indicators derived from phase 1, including the mapping of both gas stations and existing green spaces, identifying gaps, and proposing new green corridors.

3 Results

This section describes the results of the application of both SN tools, using the selection of issues, categories, and indicators presented in Table 1 and the application of the GIS-impact assessment described in Sect. 2. The discussion of these results has allowed the formulation of a number of recommendations for the strategy development, as presented in Sect. 4.1.

In this section, each indicator, selected according to the methodology described in Sect. 2.1, is illustrated in detail; an appropriate assessment methodology is explained and applied to the Bağcılar district, in Istanbul, considering the parameters indicated in Table 1 and the availability of data in the urban database.

3.1 Use of Land Biodiversity

Population Density

This topic explores the concept of population density within the context of urban structure. The intent of this analysis is to assess the proximity between residents and local goods and services by measuring population density in built-up areas, excluding green and blue spaces like parks and water bodies.

Assessment methodology:

- **Step 1**: Calculate the total population of the neighborhood. In this case, the population of Bağcılar in 2023 is **719,071 inhabitants** (represented as "A" or the numerator).
- **Step 2**: Calculate the total area of the neighborhood. Bağcılar covers **22 km^2**.
- **Step 3**: Divide the population (A) by the area (B) to get the population density.

The calculation is as follows:
Population Density = 719,071 inhabitants/22 km^2 = 32,685 inhabitants/km^2
This analysis highlights the importance of urban planning in addressing the challenges posed by high population density, as well as the opportunities for sustainable land use and improved living conditions in the area.

Availability of Green Urban Areas

This indicator focuses on **green urban areas** and the **availability of green spaces** in Bağcılar. It measures the proportion of green spaces and assesses their role in urban sustainability.

The purpose is to improve the permeability of the area (allowing water and air to pass through) and to benefit from the availability of green spaces. Green spaces are crucial as they help capture pollutants, reduce the "urban heat island" effect (where urban areas are significantly warmer due to human activity), and provide recreational spaces for the community.

The key indicator is the **proportion of vegetated areas** in the neighborhood, expressed as a percentage of the total area.

Assessment methodology:

1. **Calculate the amount of vegetated areas** in hectares (A), which is the numerator.
2. **Determine the total area of the neighborhood** (B), which serves as the denominator.
3. **Calculate the percentage** by dividing A by B and multiplying by 100 to express the proportion of green spaces in the area.

According to the **Bağcılar 2023 municipal report**, there are **176 parks and green spaces** in the district, covering a total area of **620,590 m^2**, which accounts for only **2.86% of the total area** of Bağcılar. This relatively low percentage highlights a lack of green spaces, which in turn suggests poor permeability and contributes to the urban heat island (UHI) effect in the area. Additionally, these green spaces are not evenly distributed but rather scattered across the district, which affects their accessibility but may still provide some recreational benefits.

Only **3%** of Bağcılar's total area is designated as green space. This low percentage indicates that Bağcılar has limited green urban areas, which poses challenges for environmental sustainability, urban heat reduction, and recreational space availability. It also reflects a need for better planning and expansion of green infrastructure within the district.

Green Areas in Relation to the Neighborhood Population

For the purpose of this analysis, the green areas are divided into three categories.

1. Passive green areas—forest cover
2. Soft green areas—grassland—green along the road
3. Active green areas—urban parks—neighborhood gardens—green play areas

Within the area analysis, the above-mentioned three categories were not all included in the calculation but were displayed and mentioned to understand the green fabric in the area. Soft green areas and active green areas are included in the calculation of the availability of green urban areas. However, passive green areas are not included in the calculation because they are not accessible to the public.

Assessment methodology:

1. Projection of the obtained data in the GIS software.
2. Through the "clip" geoprocessing tool, focus on the selected area.
3. Classification of the urban greenery.
4. Analyze the sqm by urban green area
5. Analyze the total number of inhabitants
6. Calculation of the availability of urban green in Bagcilar [mq/ inhabitants] (Fig. 2).

Bagcilar Green Space Area: (2024): 620.590 m^2
Bagcilar Population: 719.071 Inhabitants
Total green area in the neighborhood by total population = (620.590 / 719.071)% = 0.863 m2/Inhabitant.

Green Area Accessibility

By analyzing the accessibility of green spaces within these concentric zones, rather than using a single buffer, we can gain a layered understanding of how green amenities are distributed in relation to population centers. This approach helps us identify areas

Fig. 2 Localization of the urban green areas in the district

of deficiency and assists in the strategic planning of urban green space distribution to improve living conditions in urban areas.

200 m: This distance represents a short, easily walkable range, typically used to assess immediate accessibility to small local parks and green areas.

400 m: Recognized as a standard measure for a 5-min walk, 400 m are used to evaluate access to community-serving green spaces, reflecting daily pedestrian routes.

500 m: This distance examines accessibility for those willing to walk slightly longer, useful for understanding the reach of larger parks that attract visitors from surrounding neighborhoods.

Assessment methodology:

1. Projection of the obtained data in the GIS software.
2. Through the "clip" geoprocessing tool, focus on the selected area.

3. Classification of the urban greenery.
4. Adding road and building data obtained from OpenStreetMap to create buffer zones for walking.
5. By using the ORC tool plugin, to create 200, 400, and 500 m buffer zones.
6. Counting the total number of buildings in the area.
7. Calculation of the buildings that are inside the buffer zones.
8. Calculate the proportion of the total buildings to those in the buffer zones [%] (Fig. 3).

A: Number of buildings in each buffer zone
500 m = 20.430
400 m = 18.970
200 m = 9.500
B: Total Number of buildings in the area = 21.606

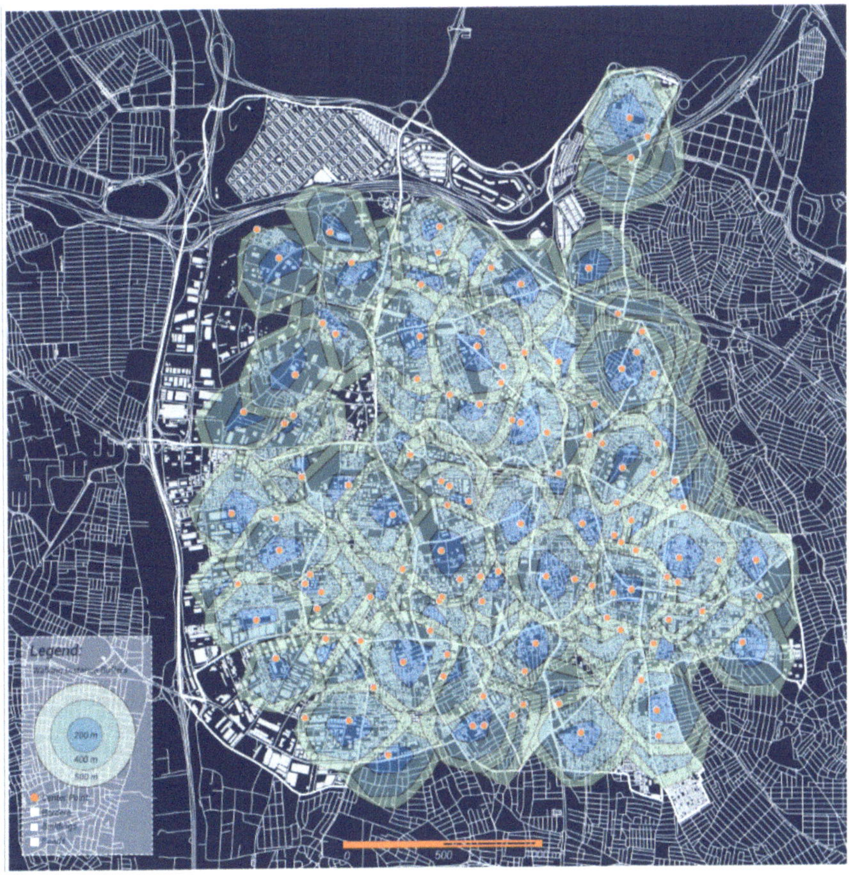

Fig. 3 Accessibility of the urban green areas in the district

A/B = %
500 m = 94.6%
400 m = 87.8%
200 m = 44.0%

3.2 Transportation and Mobility

Performance of the Public Transport System

By analyzing the accessibility of public transportation stops within these concentric zones, rather than using a single buffer, we can gain a layered understanding of how transportation stops are distributed in relation to population centers. This approach helps us identify areas of deficiency and assists in the strategic planning of urban green space distribution to improve living conditions in urban areas.

200 m: This distance is considered very convenient and represents a minimal walking effort. It is typically used to evaluate access to immediate, nearby transit options, ensuring that stops are within a quick, easy reach for daily commuters.

400 m: Often referred to as a standard walking distance in transit planning, 400 m is approximately a 5-min walk. This buffer is used to assess a reasonable walking distance for accessing more frequent or major transit services, such as busier bus lines or tram stops.

500 m: This extended distance caters to individuals who are comfortable walking a bit further, potentially enhancing access to rapid transit options like metro stations. It is useful for evaluating the catchment area of significant public transport services that serve larger urban areas.

Assessment Methodology:

1. Locate the public/municipal transport stops with daily total service frequency of at least 20trips that serve the neighborhood.
2. Through the "clip" geoprocessing tool, focus on the selected area.
3. Classification of the public transportation stops in order to analyze.
4. Adding road and building data obtained from OpenStreetMap to create buffer zones for walking.
5. By using the ORC tool plugin to create 200, 400, and 500 m buffer zones.
6. Counting the total number of buildings in the area.
7. Calculation of the buildings that are inside the buffer zones.
8. Calculate the proportion of the total buildings to those in the buffer zones [%].

A: Number of buildings in the each buffer zone
500 m = 21.563
400 m = 21.167
200 m = 13.269
B: Total Number of buildings in the area = 21.606

A/B = %
500 m = 99.8%
400 m = 98%
200 m = 61.4%

Electric Vehicle Infrastructure (Charging Stations)

In recent years, particularly within the past five years, the sales of electric vehicles (EVs) in Turkey have shown a remarkable upswing, mirroring the global trend. Unfortunately, Turkey does not have enough charging points against the remarkable use of EV cars.

Assessment methodology

1. Calculate the number of charging stations for electric vehicles. (A)- Numerator.
2. Calculate the neighborhood's population. (BJ-Denominator).
3. Calculate the indicator as: A/B.

A. Charging stations = 49 spots approximately 250 cars capacity.
B. Neighborhood's population = 745.125 Inhabitants.
    A/B = 250/719.071 = 0.00035 per inhabitants (insufficient).

Personal EV car use has rapidly increased over the years. In order to use these EV cars efficiently and increase demand, infrastructure should be well distributed in the city. According to EPDK's app where you can see bunch of brands charging stations current number of public charging stations in Istanbul are one thousand four hundred and eighty six. For the district of Bagcılar has approximately 0.00035 charging stations per inhabitant total number of 49 stations and 250 spots, indicating a critical need for expanded infrastructure to support the growing number of electric vehicles and promote sustainable transportation.

Bicycle network

Using bicycles is a sustainable and healthy transportation option for developed cities worldwide. As it was mentioned before the existence of, bicycle paths in Istanbul are only present in national parks and coastal parks. The case of the Bagcilar problem is the same as the city. The current bicycle network in Bağcılar is too sparse to calculate meaningful statistics. Such as the bicycle parking stops, as we can see in the figure that, there are only six parking spots which are not intersect with the paths. In order to emphasize the use of bicycles as a method to reduce traffic congestion and pollution, this indicator defines what does city already has.

Assessment methodology

1. Calculate the total length of bicycle paths/lanes in the neighborhood. (A)- Numerator.
2. Estimate/calculate the total number of inhabitants in the neighborhood. (B)- Denominator.
3. Calculate the indicator as: A/B.

(A) Length of bicycle paths/lanes (m) = 6600 m.
(B) Neighborhood's population = 719.071 Inhabitants.
Calculate the indicator as: A/B = 6600/719.071 = 0,0092 m/inhabitants.

Availability of Bicycle Parking Facilities

To promote cycling as an alternative to vehicle use by providing a safe and efficient mobility network.

Assessment methodology

1. Calculate the number of bicycles parking available in the neighborhood. (A)-Numerator.
2. Calculate the neighborhood's population. (BJ-Denominator).
3. Calculate the indicator as: A/B.

(A) Bicycles parking = 6 spots approximately 100 bike capacity.
(B) Neighborhood's population = 719.071 Inhabitants.
A/B = 100/719.071 = 0.00014 (insufficient).

3.3 Social Aspects

Availability and Proximity of Key Services

The accessibility and proximity of key services for local residents (e.g., schools, sport facilities, supermarkets, community buildings, etc.) are measured by percentage of inhabitants that are within a 200,400 snf 500 m of key services.

Selected key services are: Schools, sports facilities, supermarkets, community buildings, health services.

By analyzing the accessibility of key services within these concentric zones, rather than using a single buffer, we can gain a layered understanding of how services are distributed in relation to population centers.

200 m: This distance is used to evaluate very close and convenient access, ideal for ensuring that essential services are within a quick and easy walking reach for residents. 400 m: Serving as a standard benchmark, a

400-m buffer is used to assess whether key services are accessible within a typical 5-min walk, supporting daily needs and activities.

500 m: This extended range is considered for evaluating access to services that residents might be willing to travel slightly longer distances to utilize, enhancing service availability across a broader area.

Assessment methodology

1. Locate the service areas with significant daily usage, such as schools, sports facilities, supermarkets, community buildings, and health services that are crucial for the neighborhood.

2. Through the “clip” geoprocessing tool, focus on the selected area containing these services.
3. Classify the service areas in order to analyze their types and functions within the community.
4. Adding road and building data obtained from OpenStreetMap to create buffer zones for walking.
5. By using the ORC tool plugin to create 200, 400, and 500 m buffer zones.
6. Counting the total number of buildings in the area.
7. Calculation of the buildings that are inside the buffer zones
8. Calculate the proportion of the total buildings to those in the buffer zones [%].

The SES index is based on a multicriterion scoring system. Using official statistics from various institutions and organizations (https://data.ibb.gov.tr/en/dataset/mahallem-istanbul-projesi-ses-sosyo-ekonomik-statu-skorlari), a SES score was calculated for each neighborhood. According to this index, two basic criteria were used to understand the socioeconomic levels of individuals, i.e., education level (E) and occupation/professional variables (P).

By the SES index, eight groups were identified as A+, A, B+, B, C+, C, D, and E. According to the multicriteria evaluation, the neighborhoods with the highest SES index are at the A+ and A levels, while the neighborhoods with the lowest SES index are at the D and E levels.

3.4 *Environmental Quality*

Air Quality

Particulate matter (PM10) concentration: 207 Days/2023.

The days with PM10 levels exceeding the daily 50 $\mu g/m^3$ limit are analyzed. According to EU standards, this should happen at most 35 days a year. In Bağcılar’s case, the data shows 57 days classified as ‘poor,’ which likely correlates with days where PM10 levels exceeded the EU limit. While the specific annual average concentration was previously provided, Bagcilar’s average 2023 PM10 level is 38.11 $\mu g/m^3$, which increased by 19.56% compared to the previous year. The current average level is about to pass the EU limit of 40 $\mu g/m^3$. To assess the long-term ambient air quality with respect to particulates < 10mu (PM10) in the neighborhood.

Assessment methodology

1. Daily test air samples in accordance with national or regional procedures over a period time of one year.
2. Evaluate the number of days exceeding the daily limits in a year.
   64 Days Exceed the limit.

4 Discussion

A discussion of the results obtained from the application of the SN tool and GIS tool to Bağcılar district of Istanbul City is provided in this section, according to the different categories: Green areas, transportation and mobility, social aspects, socioeconomic status, and air quality.

Green areas

The uncontrolled construction process in Bağcılar district, marked by a lack of planning, projects, licenses, inspections, and proper settlements, has led to a significant shortage of reinforcement areas. Contrary to what its name might suggest, Bağcılar is extremely deficient in green spaces. The district offers a total of 0.620 square kilometers (620,000 square meters) of green space. Per capita, this translates to only 0.863 square meters of green space which is insufficient to meet the needs of the population and contributes to environmental issues such as the urban heat island effect and poor air quality. Increasing green spaces is essential to improve environmental quality and provide recreational opportunities for residents. The district includes a total of 176 parks and green areas, including three special theme parks. In rethinking the development of Bağcılar district, addressing the need for green spaces should be a priority, and it should be done in a systematic and planned manner, much like other essential infrastructure improvements.

Transportation and mobility

Bağcılar district, strategically located between the Trans-European Motorway (TEM) and E5 main transportation roads, offers convenient access from surrounding areas, even during heavy Istanbul traffic. However, the unplanned development within the district has made transportation in the center highly problematic. The roads within Bağcılar urgently need to be upgraded to meet modern standards. Reorganizing the urban roads in Bağcılar district is crucial for addressing the transportation needs of both residents and those coming from outside the district to access public and private sector services, especially hospitals. The inefficiencies in the district's transportation network directly impact the accessibility and quality of health services, notably at the Bağcılar State Hospital, a large-scale research facility. The congestion caused by this health complex underscores the need for better road infrastructure. Despite these challenges, Bağcılar remains an attractive area due to its proximity to major highways and transportation routes. This accessibility makes it a preferred location for businesses seeking operational convenience. Ensuring that the district's internal transportation network matches the ease of access provided by surrounding roads is essential to maintain its appeal and functionality. According to the mobility map and public transportation performance analysis, Bagcilar performed well. The wide rubber transportation is supported by a rail transportation system that crosses the area from the middle. The buffer performance analysis shows us approximately 99.8% of buildings are within 500 m of a public transportation stop, 98% are within 400 m, and 61.4% are within 200 m. However, the service frequency and efficiency vary, necessitating improvements to ensure reliable and convenient access for all residents. Enhanced public transportation can significantly reduce traffic congestion and improve mobility across the district.

The bike network in the area is notably inadequate, with only 6.6 km of bike paths and just 100 official bike parking spaces. This limited infrastructure makes it challenging for residents who prefer cycling as a mode of transportation. Enhancing the bike network is essential to promote sustainable mobility and provide a viable alternative to car travel, contributing to a healthier and more environmentally friendly community. Although the availability of EV charging stations has increased rapidly in Istanbul, the area still faces significant challenges.

With only 250 charging stations, the ratio equates to a mere 0.00035 charging stations per inhabitant. This extremely low coverage makes it difficult for electric vehicle users to find convenient charging options, highlighting the urgent need for expanding the EV charging infrastructure to support the growing number of electric vehicle users and promote sustainable transportation. Expanding sustainable transport options, such as bicycle paths and EV charging infrastructure, is crucial to promoting environmentally friendly mobility solutions and reducing the district's carbon footprint.

Social aspects

Land use of the map and the service area analysis show us that residential areas form the area widely. General service area performance distribution indicates that Mahmutbey and Goztepe have the lowest service availability in the area. Although accessibility to healthcare facilities within 200, 400, and 500-m buffers is generally adequate, with a coverage rate of 91.5% within a 500-m buffer zone.

This ensures that a vast majority of the population has convenient access to health services, promoting overall well-being and timely medical care. However, the increasing population places immense pressure on these services. Expanding healthcare infrastructure and improving service delivery are critical to meeting the growing demand and ensuring residents have access to quality healthcare [28].

Access to supermarkets is relatively good, with 56.2% of buildings within 200 m, 83.9% within 400 m, and 89.9% within 500 m. This widespread availability ensures that nearly all residents can conveniently access supermarkets, facilitating easy procurement of daily necessities.

Bağcılar has 84 educational institutions, including 3 kindergartens, 55 primary schools, 15 high schools, and 7 vocational high schools. In contrast, in the Bagcilar district, schools cover approximately 50% of the population within a 500-m buffer zone. While this indicates a moderate level of accessibility, there is significant room for improvement to ensure better educational access for the community.

Access to sports facilities is limited, with 7.7% of buildings within 200 m, 23.6% within 400 m, and 31.1% within 500 m. According to a recent survey, residents have expressed concerns about the lack of sports facilities, which negatively impacts their health and well-being. Addressing this shortfall is crucial for promoting a healthier and more active community.

The performance of community buildings is even more concerning, with only 16.1% of the population having access within a 500-m buffer zone. This low coverage rate highlights a critical need for improved communal support infrastructure. Enhancing these facilities is essential to meet the area's socioeconomic needs and foster a stronger sense of community.

Socioeconomic status

Bağcılar's socioeconomic index reveals substantial disparities among neighborhoods, primarily due to low educational attainment and limited professional opportunities. Addressing these disparities requires focused efforts to improve education and vocational training, thereby enhancing the socioeconomic status of residents. Programs aimed at economic development and poverty alleviation are crucial for promoting social equity and economic stability.

Air quality

In 2023, Bağcılar experienced 57 days classified as 'poor,' likely due to PM10 levels exceeding the EU limit. The average PM10 level for the year was 38.11 $\mu g/m^3$, a 19.56% increase from the previous year. This current average is approaching the EU limit of 40 $\mu g/m^3$, indicating a worsening air quality trend. The main reasons for poor air quality include high traffic emissions, industrial activities, dust from ongoing construction projects, insufficient green spaces, and the use of coal and other solid fuels for heating. Addressing these pollution sources is crucial to improve air quality, protect public health, and enhance the quality of life for residents in Bağcılar.

4.1 *Strategic Recommendations*

The results discussed in previous section have highlighted a number of critical areas to be addressed with targeted strategies by the city planning for overcoming Bağcılar current challenges and pave the way for sustainable development, ensuring a higher quality of life for its residents and a more resilient and sustainable urban environment [23, 30]. The strategies and targets are illustrated in Fig. 4. These strategic recommendations are presented as follows:

Urban planning and infrastructure development: Implement comprehensive urban planning strategies to manage population density and ensure adequate infrastructure, including roads, public transportation, and green spaces.

Sustainable transportation initiatives: Develop and promote sustainable transportation options, including expanding the network of bicycle paths and increasing the availability of electric vehicle charging stations.

Enhanced public services: Improve the quality and accessibility of education to raise literacy rates and educational attainment. Additionally, enhance the distribution and quality of key public services such as health facilities, sports centers, and community buildings to ensure they are accessible to a larger proportion of the population.

Expansion of green spaces: Increase the availability of green urban areas to at least meet the recommended standards of green space per inhabitant. This can be achieved

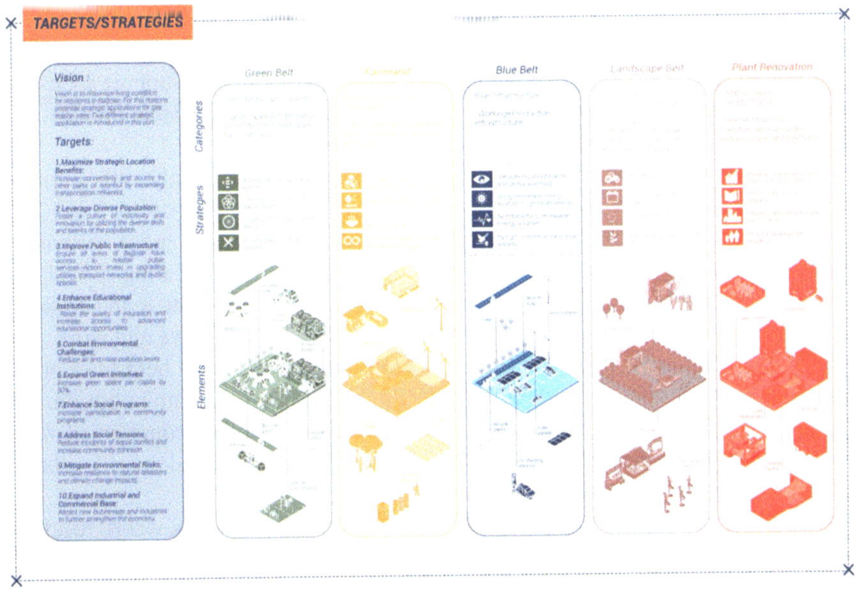

Fig. 4 Strategies and targets for the Bağcılar district

through the development of new parks and the integration of green infrastructure in urban planning.

Improved socioeconomic development: Implement programs to address socioeconomic disparities by improving education and vocational training, promoting economic development, and providing targeted support to low-income neighborhoods.

Environmental protection measures: Implement measures to reduce air pollution and improve environmental quality, such as increasing green spaces, regulating industrial emissions, and promoting clean energy solutions.

5 Conclusions and Future Developments

This study has recognized the need for urban regeneration in cities like Istanbul, focusing on detailed analysis at district level. Bağcılar has been considered as pilot case study. This faces significant challenges and opportunities due to its rapid urbanization and population growth. This study has developed and experimented with a new SN tool-based spatialized (GIS-based) framework with the aim of repurposing underutilized gas station sites into green areas that enhance the urban landscape, improve environmental quality, and promote community resilience.

The comprehensive methodology and carefully selected indicators have provided a solid framework for planning and implementing these transformations, drawing on successful international examples and local context. Results have shown that repurposing gas stations in Bağcılar can contribute to more efficient land use, providing essential amenities in densely populated areas. This strategy can support socioeconomic development by creating new public spaces, enhancing local amenities, and providing job opportunities in sustainable industries.

Bağcılar's conversion of gas stations into green spaces can significantly improve the district's livability by:

Enhancing green coverage: Increasing the per capita green space, providing residents with more opportunities for outdoor activities, and creating a healthier living environment. The creation of green spaces can also contribute to the aesthetic enhancement of the district, making it a more attractive place to live and work.

Reducing pollution: Plants and trees in these green spaces will help filter pollutants, improving air quality and reducing health risks associated with pollution. The introduction of green infrastructure can also contribute to the reduction of noise pollution. The combined effect of improved air quality and reduced noise levels can enhance overall public health and well-being.

Fostering community: Green spaces can act as communal areas where residents can gather, interact, and build stronger community ties, enhancing social cohesion. Community gardens and urban farms can provide opportunities for local food production and community engagement. These spaces can also host cultural and social events, fostering a sense of community and belonging residents.

Community involvement: Engaging local residents in the planning and upkeep of these green spaces ensures that they are utilized effectively and appreciated by the community. Volunteer programs and community stewardship initiatives can foster a sense of ownership and responsibility. Regular community feedback sessions can help identify areas for improvement and ensure that the spaces remain relevant and useful.

Policy support: Strong support from local and city-wide policies will be essential in maintaining and expanding green infrastructure initiatives. Policy frameworks should incorporate green infrastructure planning into zoning regulations and development codes. Incentives for developers to include green spaces in their projects can also promote widespread adoption of green infrastructure principles.

Adaptive management: Monitoring and adapting strategies based on feedback and changing environmental conditions will help sustain the benefits of these green spaces. Regular assessments and updates to management practices can ensure the long-term success of green infrastructure projects. The use of technology, such as remote sensing and environmental sensors, can aid in the monitoring and management of these spaces.

In conclusion, the repurposing of gas stations in Istanbul, particularly in Bağcılar, represents a strategic approach to urban regeneration. By transforming these underutilized spaces into green areas, the city can address critical urban challenges such as environmental sustainability, public health, and community resilience.

By prioritizing the development of green infrastructure, cities can enhance livability and sustainability, employment opportunities in sustainable industries, improving the socioeconomic ensuring a better quality of life for current and future generations. The comprehensive approach to urban regeneration presented in this study provides a replicable framework that can be adapted to different urban contexts globally. The future scenarios outlined in this study offer a vision of an inclusive, resilient, and sustainable urban future, setting a benchmark for innovative urban regeneration practices worldwide.

References

1. Abbott C (2002) Portland's Willamette River Greenway: new uses for old rivers. Urban Design International 7(4):235–245
2. Adams D, Tiesdell S (2013) Shaping places: urban planning, design and development. Routledge
3. Akın A, Sunar F, Berberoğlu S (2015) Urban change analysis and future growth of Istanbul. Environ Monit Assess 187:506
4. Alberti M (2005) The effects of urban patterns on ecosystem function. Int Reg Sci Rev 28(2):168–192
5. Bağcılar Municipality (2020) Bağcılar Municipality Report. Bağcılar Municipality
6. Bulkeley H, Betsill MM (2005) Rethinking sustainable cities: Multilevel governance and the 'urban' politics of climate change. Environ Politics 14(1):42–63
7. Chiesura A (2004) The role of urban parks for the sustainable city. Landsc Urban Plan 68(1):129–138
8. City of New York (2015) One NYC: the plan for a strong and just city. New York City Mayor's Office
9. City of Vienna (2014) Urban development plan Vienna (STEP 2025). City of Vienna
10. Demir O, Yılmaz S (2017) Green space policy in Istanbul. Urban Forestry Urban Greening 25:76–83
11. Ferrell J (2019) The gas station in the urban landscape. Urban Stud 56(10):2075–2091
12. Gencer EA (2013) The making of a disaster: Earthquake Hazard and Urban Development in Istanbul. In: The interplay between urban development, vulnerability, and risk management. SpringerBriefs in Environment, Security, Development and Peace, vol 7. Springer, Heidelberg
13. Gilardi R (ed) (2020) Low carbon and urban morphologies. Progetto Europeo Moloc-Divisione Urbanistica e Territorio Area Urbanistica Qualità dell'Ambiente Costruito. Proposta Tecnica Del Progetto Preliminare (Ai Sensi Dell'artt. 14 E 15 Della Lur N. 56/1977 E Smi) Documentazione Di Studio, Quaderno 6/2. Torino, Maggio 2020. Available at: http://geoportale.comune.torino.it/web/sites/default/files/mediafiles/4.9_quaderno_6-2_relazione_moloc.pdf
14. Greater London Authority (2021) The London Plan: the spatial development strategy for Greater London. Greater London Authority
15. Günay Z (2016) The role of urban design in cultural sustainability: the case of Istanbul. J Urban Des 21(1):82–100
16. Heywood I, Cornelius, S, Carver S (2011) An introduction to geographical information systems. Pearson Education
17. Istanbul Metropolitan Municipality (2022) Istanbul vision 2050 strategy document. Retrieved from Istanbul Municipality
18. Kim J (2012) Seoul's Cheonggyecheon restoration project: a critical review. Pac Aff 85(2):355–369
19. Larsson N (2019) iiSBE SNTool for performance assessment of urban neighbourhoods ! iiSBE (May). https://doi.org/10.13140/RG.2.2.32305.84326
20. Lombardi P et al (2017) Multicriteria spatial decision support systems for future urban energy retrofitting scenarios. Sustainability (Switzerland) 9(7). https://doi.org/10.3390/su9071252

21. Nakamura K (2018) Sustainable urban regeneration: the case of Tokyo's repurposed gas stations. Sustain Cities Soc 39:245–254
22. Pignatelli M, Torabi Moghadam S, Genta C, Lombardi P (2023) Spatial decision support system for low-carbon sustainable cities development: an interactive storytelling dashboard for the city of Turin. Sustain Cities Soc 89
23. Pinder D (2011) City at the crossroads: Istanbul and the struggle for urban regeneration. Int J Urban Reg Res 35(3):523–536
24. Plaza B (2000) Evaluating the influence of a large cultural artifact on tourism development: the case of the Guggenheim Museum Bilbao. Urban Aff Rev 36(2):264–274
25. Rohracher H, Späth P (2014) The interplay of urban energy policy and sociotechnical transitions: the eco-cities of Graz and Freiburg in retrospect. Urban Stud 51(7):1415–1431
26. SNTool MED (2023) Sustainability assessment tool for Mediterranean neighborhoods. European Union's ENI CBC MED Programme. Available at: https://sustainablemedcities.tools/manuals/Manual_SNTool_MED_Cities.pdf?v=20231018
27. Stahre P (2008) Blue-green stormwater management practices: planning and implementation in the city of Malmö, Sweden. Vatten 64(1):291–302
28. Sönmez Özdemir S, Erkut G (2024) Drivers of socio-spatial change in Istanbul: historical and longitudinal analysis of 5 cases from Bakırköy. J Des Resilience Archit Plann 5(1):96–122
29. Torabi MS, Delmastro C, Corgnati S, Lombardi P (2017) Urban energy planning procedure for sustainable development in the built environment: a review of available spatial approaches. J Clean Prod 165:811–827
30. Wolfram M (2016) Conceptualizing urban transformative capacity: a framework for research and policy. Cities 51:121–130
31. Woodcock I, Dovey K, Wollan S (2011) Speculating on Melbourne's urban design futures. Urban Policy Res 29(3):251–265
32. Yılmaz S (2020) Karaköy Gas Station Park: a proposal for urban green space in Istanbul. Urban Forestry Urban Greening 47:126539
33. Özdemir S (2021) Taksim Charging Hub: integrating green infrastructure in Istanbul. Renew Energy 162:762–771

Future of Net Zero Carbon Practices in Construction

N. Domingo and A. Le

1 Introduction

In the concluding chapter of *Global Net Zero Carbon Practices in Construction*, the authors synthesise the book's central insights, presenting an in-depth perspective on the insights, obstacles, and future directions for achieving net zero carbon (NZC) within the construction industry. Over recent decades, NZC has evolved from scientific concepts in physics and chemistry to a central strategy in global greenhouse gas reduction, underpinned by initiatives like the Kyoto Protocol and the Paris Agreement. This shift has prompted the construction sector to recognise its crucial role in NZC through advancements in building design, lifecycle assessments, renewable energy integration, and reduced embodied carbon.

2 Insights into Net Zero Carbon Building Design and Construction

This book highlights the growing importance of NZC concepts, including net zero carbon buildings (NZCBs) and net zero energy buildings (NZEBs), which aim to balance carbon emissions across a building's entire lifecycle. Despite significant progress, the lack of global standardisation has led to fragmented approaches, underscoring the need for unified definitions and consistent metrics.

The evolution of NZC initiatives can be categorised into three primary phases. The initial phase, influenced by the Kyoto Protocol (2005–2012), focused on carbon neutrality through energy efficiency and offset strategies. The subsequent pre-Paris Agreement phase (2012–2015) saw heightened carbon reduction ambitions,

N. Domingo (✉) · A. Le
School of Built Environment, Massey University, Auckland, New Zealand
e-mail: n.d.domingo@massey.ac.nz

N. Domingo et al. (eds.), *Global Net Zero Carbon Practices in Construction*, Green Energy and Technology, https://doi.org/10.1007/978-981-96-3853-6_11

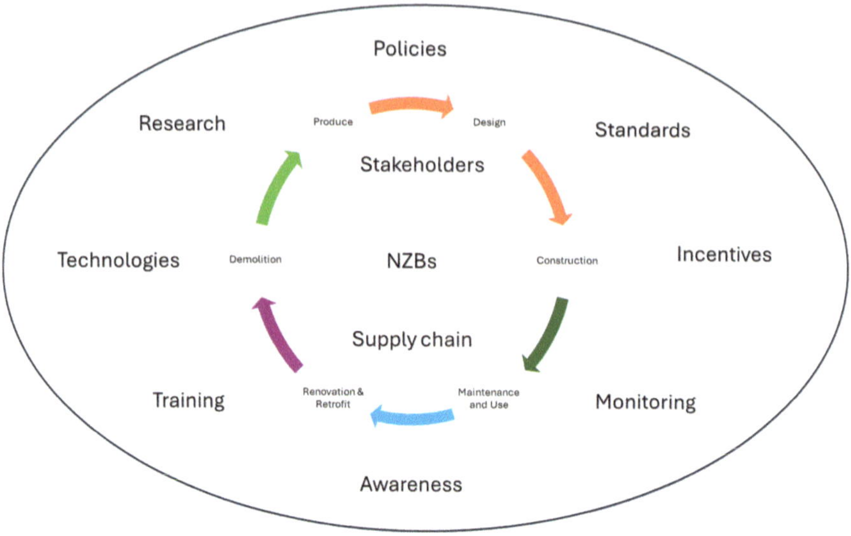

Fig. 1 Framework for NZCBs through lifecycle approach

setting the foundation for the post-Paris Agreement phase (2015–2024), which has brought about major NZC commitments, technological advancements, and widespread industry shifts. However, the book emphasises that meeting NZC goals requires a transformative, lifecycle approach at all stages of construction from raw material extraction to eventual demolition (Fig. 1) to reduce both operational and embodied carbon.

The global construction sector contributes significantly to greenhouse gas (GHG) emissions, with buildings accounting for nearly 39% of global energy-related carbon emissions. Achieving NZC in this sector requires addressing both operational and embodied carbon. Operational carbon results from energy consumption over a building's lifespan, while embodied carbon accumulates during the material extraction, production, and construction stages. Tackling both is essential to meet global emissions reduction targets, particularly the goal of net zero by 2050 as outlined in the Paris Climate Agreement, which has motivated governments, industries, and stakeholders worldwide to adopt low-carbon practices.

2.1 Definitions

A major challenge in achieving NZC in construction is the inconsistency in definitions and standards across regions. Countries like the UK, New Zealand, and

Australia have established comprehensive frameworks to standardise NZC practices, whereas other regions maintain narrower definitions, limiting their application. This disparity creates a patchwork of standards, making it difficult to implement universally effective NZC strategies. Developed countries are advancing clean energy technologies, while developing countries face obstacles due to limited financial resources and policy support, despite notable efforts in green infrastructure and low-emission standards. To promote NZC construction, efforts must consider these varied regional landscapes, building on successful examples of multi-stakeholder engagement.

2.2 Collaboration and Stakeholder Engagement

Collaboration across all levels of the construction industry including policymakers, professionals, end-users, and civil society is essential to advancing NZC goals. Multi-stakeholder engagement fosters knowledge sharing, aligns best practices, and helps address common barriers such as fragmented regulations and limited incentives. Actively involving stakeholders throughout a building's lifecycle promotes broader and more effective adoption of NZC practices.

Educational institutions also play a crucial role, equipping the workforce with skills in low-carbon construction techniques, carbon tracking, and lifecycle assessments. Real-world training opportunities bridge the gap between academic knowledge and practical application, empowering future professionals to support the NZC transition effectively.

2.3 Technological Innovations for Carbon Reduction

Technological advancements are transforming the industry's ability to track, manage, and reduce carbon emissions. Tools such as Building Information Modelling (BIM), carbon accounting software, and AI-driven analytics enable real-time emissions tracking, predictive maintenance, and informed decision-making across project lifecycles. These technologies support the management of both operational and embodied carbon, though greater integration is needed for widespread adoption, particularly in regions with limited financial resources. Continued research is vital to optimise these technologies, making them accessible across diverse geographies and building lifecycle stages.

2.4 Lifecycle Approach to Carbon Reduction

Achieving NZC in construction demands a holistic, lifecycle approach to carbon reduction. Operational emissions can be minimised through energy-efficient designs, passive strategies, and renewable energy integration. Reducing embodied carbon, meanwhile, requires a shift towards low-carbon materials and construction methods, such as modular and prefabricated designs. Passive design elements like optimised orientation and insulation, combined with active solutions like energy-efficient HVAC systems and renewable energy sources, help reduce a building's overall carbon footprint. Retrofitting existing structures with energy-efficient systems offers one of the most immediate and impactful ways to cut emissions.

2.5 Economic and Environmental Evaluations

Economic viability is essential to promoting sustainable practices across the sector. Comprehensive evaluations such as net present value (NPV), return on investment (ROI), and the carbon reduction efficiency (CRE) metric help building owners and managers justify investments in low-carbon technologies. These assessments allow for a clearer understanding of financial savings from energy efficiency improvements alongside carbon reduction benefits. As green financing options and government incentives expand, economic barriers to sustainable retrofitting are gradually being lowered, making it more accessible to a broader range of stakeholders.

2.6 Addressing Construction and Demolition Waste (CDW)

Managing CDW is another significant challenge in achieving NZC. CDW generates environmental impacts, including greenhouse gas emissions and resource depletion. Moving towards a circular economy in construction by focusing on material reuse, recycling, and waste reduction can significantly mitigate these effects. Advanced technologies, such as automated waste sorting and BIM, improve waste management efficiency; however, their adoption is limited by high costs and technical expertise requirements. Governments need to continue supporting and incentivising the industry to adopt sustainable waste management practices.

3 Barriers to NZCB Implementation

Achieving NZC in the construction industry is hindered by several barriers, including regulatory inconsistencies, financial limitations, and technological gaps. In developed countries with legally binding NZC goals, robust policy frameworks have facilitated progress yet delays in regulatory updates and insufficient monitoring present ongoing challenges. In contrast, countries with non-binding commitments, often developing nations, encounter significant difficulties due to high initial costs and limited access to low-carbon technologies. The industry also faces substantial challenges with high costs of green technologies, supply chain limitations, and a scarcity of skilled personnel necessary to implement NZC projects effectively.

A further barrier lies in existing buildings, which contribute significantly to global emissions due to both their operational inefficiencies and embodied carbon. Policies have primarily focused on new construction, while the need for large-scale retrofitting of older structures remains under-addressed. Effective NZC strategies must incorporate holistic retrofitting solutions and advanced lifecycle assessment tools to reduce both operational and embodied emissions.

4 Opportunities to NZCB Implementation

A pathway forward to NZC involves enhancing collaboration across governmental, industry, and research sectors to bridge regulatory gaps and foster a unified understanding of NZC objectives. The book underscores the importance of stakeholder engagement, as NZC achievements depend on coordination between policymakers, architects, contractors, investors, and end-users. Improved data sharing, mandatory reporting, and enhanced communication can foster transparency and trust across the industry.

During the design phase, integrating both passive and active design strategies such as natural lighting, renewable energy adoption, and energy-efficient HVAC systems can substantially lower carbon emissions. Yet, challenges persist, especially in balancing environmental goals with practical construction limitations. Innovative tools like BIM play a critical role, supporting lifecycle assessments and helping stakeholders make informed design choices.

5 The Path Forward for Net Zero Carbon Buildings

Achieving NZCBs on a global scale requires a cohesive, lifecycle-based approach supported by aligned regulatory frameworks, robust data systems, and financial incentives. This approach must encompass all stages of the building lifecycle from design and construction through to demolition. Coordinated international efforts will enable industry stakeholders to adopt sustainable practices more effectively. By advancing circular economy models, implementing NZC-aligned policies for both new and existing buildings, and fostering active stakeholder engagement, the construction industry can play a pivotal role in meeting global NZC targets. Below are key areas of focus to support this transition:

- **Policy and technological implementation**: Governments must embed NZCB strategies within their climate action plans, aligning with Paris Agreement targets. Scaling successful case studies and implementing these strategies across the industry will drive substantial progress.
- **Standardise definitions**: Establishing consistent global definitions for NZCBs is essential to ensure alignment in carbon neutrality efforts, allowing for streamlined policies, shared goals, and clear targets across borders.
- **Strengthen international collaboration**: Partnerships with nations that lead in NZCB practices, such as the UK, USA, China, Canada, and New Zealand, are essential for knowledge exchange, policy-sharing, and developing global standards to harmonise best practices across borders.
- **Track future trends**: Continuous monitoring of NZCB-related innovations will help the industry stay adaptable to new technologies, further supporting carbon reduction efforts in construction.
- **Emphasise a lifecycle focus**: Industry practices should prioritise carbon reduction across the entire building lifecycle, including embodied carbon during construction, operational carbon during use, and sustainable disposal practices.
- **Promote carbon removal technologies**: Embracing carbon removal innovations within construction processes will help create carbon-negative buildings, significantly contributing to NZC targets.
- **Strengthen policy alignment**: Local policies should align with global frameworks like the Paris Agreement, addressing both operational and embodied carbon to create a unified approach to carbon reduction.
- **Enhance global collaboration**: Increased international collaboration, particularly with countries with advanced research and regulatory frameworks, is critical to accelerate progress on NZCBs through shared strategies and pooled expertise.
- **Expand NZCB research**: Future research should address emerging challenges and technologies, with a focus on lifecycle carbon reduction, carbon estimation tools, and practical strategies for reducing both operational and embodied carbon.

6 Conclusion

The insights from this book serve as both a call to action and a roadmap for the construction industry's transformative journey towards a sustainable, net zero carbon future. By prioritising these focus areas, the global construction sector can lead the way in climate action, setting a powerful example of how aligned strategies, innovative technologies, and proactive policies can turn the vision of net zero carbon buildings into reality.

MIX
Papier aus verantwortungsvollen Quellen
Paper from responsible sources
FSC® C105338

If you have any concerns about our products,
you can contact us on
ProductSafety@springernature.com

In case Publisher is established outside the EU,
the EU authorized representative is:
Springer Nature Customer Service Center GmbH
Europaplatz 3, 69115 Heidelberg, Germany

Printed by Libri Plureos GmbH
in Hamburg, Germany